DON'T HURT PEOPLE AND DON'T TAKE THEIR STUFF

Also by Matt Kibbe

Hostile Takeover
Give Us Liberty (coauthor)

DON'T HURT PEOPLE

AND DON'T TAKE THEIR STUFF

A LIBERTARIAN MANIFESTO

MATT KIBBE

WM

WILLIAM MORROW

An Imprint of HarperCollins*Publishers*

For Terry, who's always there for me

HarperCollins books may be purchased for educational, business, or sales promotional use. For information please e-mail the Special Markets Department at SPsales@harpercollins.com.

A hardcover edition of this book was published in 2014 by William Morrow, an imprint of HarperCollins Publishers.

FIRST WILLIAM MORROW PAPERBACK EDITION PUBLISHED 2015.

Library of Congress Cataloging-in-Publication Data has been applied for.

ISBN 978-0-06-230827-6

15 16 17 18 19   DIX/RRD   10 9 8 7 6 5 4 3 2 1

In the United States, where it has become almost impossible to use "liberal" in the sense in which I have used it, the term "libertarian" has been used instead. It may be the answer; but for my part I find it singularly unattractive. For my taste it carries too much the flavor of a manufactured term and of a substitute. What I should want is a word which describes the party of life, the party that favors free growth and spontaneous evolution. But I have racked my brain unsuccessfully to find a descriptive term which commends itself.

—FRIEDRICH HAYEK, "WHY I AM NOT A CONSERVATIVE"

CONTENTS

CHAPTER 1

RULES FOR LIBERTY

DON'T HURT PEOPLE, AND don't take their stuff. That's it, in a nutshell. Everyone should be free to live their lives as they think best, free from meddling by politicians and government bureaucrats, as long as they don't hurt other people, or take other people's stuff.

I believe in liberty, so the rules are pretty straightforward: simple, blindly applied like Lady Justice would, across the board. No assembly required.

To me, the values of liberty just seem like a common-sense way to think about political philosophy. The rules are easily understood, our aspirations for government are modest and practical, and our designs on the lives and behavior of others are unpresumptuous, even humble.

There is a renewed and heated debate about the future of America going on right now. Our government seems broken. What is the best way to get our mutually beloved country back on track? People are seeking answers. When you get past all the acrimony and all the name-calling, the question we are all debating is really quite simple: Do you believe in the freedom of individuals to determine their own futures and solve problems cooperatively working together, or do you believe that a powerful but benevolent government can and should rearrange outcomes and make things better?

More and more, the debate about how we live our lives and what the government's legitimate role is in overruling our personal decisions has become increasingly polarized, even hostile. The president is fighting with Congress. Democrats are fighting with Republicans. Conservatives are fighting with liberals. Libertarians are fighting with "neo-cons." Political insiders and career bureaucrats are pushing back against the wishes of grassroots Americans. And left-wing "progressives" are attacking, with increased vitriol, tea party "anarchists." It's enough to make your head spin, or at least make you rationally opt out of the whole debate as it is defined by all of the experts that congregate in Washington, D.C., or on the editorial pages of the most venerated newssheets of record.

Normal people—real Americans outside the Beltway—have better things to do. They should focus on their lives and their kids and their careers, their passions and their goals and their communities. Right?

Except that we just can't anymore. It seems like the decisions Washington power brokers make about what to do for us, or to us, or even against us, are having an increasingly adverse impact on our lives. Young people can't find jobs, and can't afford to pay off their student loans. Parents are having an increasingly hard time providing for their families. Seniors can't afford to retire, and their life savings seem to be shrinking for reasons that are not quite clear. And every one of us is somehow being targeted, monitored, snooped on, conscripted, induced, taxed, subsidized, or otherwise manipulated by someone else's agenda, based on someone else's decisions, made in some secret meeting or by some closed-door legislative deal in Washington, D.C.

What gives, you ask?

It seems like we have reached a tipping point where governance in Washington and your unalienable right to do what you think best for yourself and your family have collided. You and I will have to get involved, to figure out what exactly the rules are, and to set them right again.

THERE ARE RULES

I am not a moral philosopher and I don't particularly aspire to be one. That said, I have stayed at more than one Holiday Inn Express. That makes me at least smart enough to know what I don't know. So the rules that follow represent my humble attempt to boil down and mash up all the best thinking in all of human history on individualism and civil society, the entire canon of Judeo-Christian teachings, hundreds of years of English Whig, Scottish Enlightenment, and classical liberal political philosophy, way too much Friedrich Hayek and Adam Smith, a smattering of karma and Ayn Rand, and, if my editor doesn't excise it out of the manuscript, at least a few subliminal hat tips to *The Big Lebowski*. All of this in six convenient "Rules for Liberty."

What on earth am I thinking? My inspiration, in an odd way, is Saul Alinsky, the famous community organizer who was so influential on two of his fellow Chicagoans—Barack Obama and Hillary Clinton. Everybody's favorite leftist famously wrote thirteen *Rules for Radicals* for his disciples to follow. His book is "a pragmatic primer for realistic radicals" seeking to take over the world.

Alinsky actually dedicates his book to Lucifer. I'm not kidding.

> *Lest we forget at least an over the shoulder acknowledgment to the very first radical: from all our legends, mythology and history (and who is to know where mythology leaves off and history begins—or which is which), the very first radical known to man who rebelled against the establishment and did it so effectively that he at least won his own kingdom—Lucifer.*

What the hell was he thinking? Just for fun, Google "Alinsky" and "Lucifer" sometime and see for yourself the rhetorical knots his admirers tie themselves into trying to explain the dedication to their favorite book, penned by their cherished mentor. Did Alinsky really mean it? Who knows, but tongue-in-cheek or not, it seems to reflect the by-any-means-necessary spirit of the book.

So, how could I find inspiration here? It's no secret that

many of us liberty-minded "community organizers" have expropriated some of Alinsky's tactical thinking in the defense of individual freedom. But I think there's a categorical difference between us and them. *Rules for Radicals* is not a tome about principles; it is a book about winning, sometimes with wickedly cynical and manipulative tactics. The principles seem to be missing, or an afterthought, something to be figured out later, air-dropped into the plan depending upon who ends up in charge. This cart-before-the-horse thinking seems to be consistent with the progressive mind-set. The rule of man instead of the rule of law, or the writing of a blank check for government agents empowered with great discretionary authority over your life. If we just suspend our disbelief and trust them, everything is supposed to turn out fine. Better, in fact.

We, on the other hand, start from first principles. The nice thing about the Rules for Liberty is that our values define our tactics, so there's no ends-justify-the-means hypocrisy. Liberty is right. Liberty is the basis for social cooperation and voluntary organizing. Liberty allows each of us to achieve what we might of our lives.

Liberty is good policy, and good politics. But good politics is a consequence, not the goal. "Liberty is not a means to a higher political end," wrote Lord Acton. "It is itself the highest political end. It is not for the sake of a good public administration that it is required, but for the security in the pursuit of the highest objects of civil society, and of private life." [1]

It's common sense. The Rules for Liberty are applied

equally, without bias or discrimination, and don't allow the moving of goalposts midgame. These rules don't permit gray-suited middlemen to rearrange things for your special benefit, or against your personal preferences, arbitrarily.

Adam Smith, the Scottish moral philosopher widely considered the father of modern economics, based his economic thinking on the mutually beneficial gains achieved from voluntary cooperation. But cooperation and exchange are based on mutually understood values. His most important work, a foundation for all classical liberal thinking, is *The Theory of Moral Sentiments*. In my book *Hostile Takeover*, I briefly discuss Smith's influence on the work of Nobel laureate economist Vernon Smith, his inquiries into the ways that the rules of community conduct function in real life. The rules that allow for peaceful cooperation emerge, seemingly spontaneously, from human actions.

How do such social norms—the rules—emerge? The question is one that F. A. Hayek, also a Nobel laureate, spent the latter half of his professional career exploring. Both Vernon Smith and Hayek find the basis for their inquiry in Smith's *Moral Sentiments:*

> *The most sacred laws of justice, therefore, those whose violation seems to call loudest for vengeance and punishment, are the laws which guard the life and person of our neighbor; the next are those which guard his property and possessions; and last of all come those which guard what are called his personal rights, or what is due to him from the promises of others.*

1. Don't Hurt People

This first rule seems simple enough, and no decent person would hurt another unless the action was provoked or in some way justified. Free people just want to be left alone, not hassled or harmed by someone else with an agenda or designs over their life and property. We would certainly strike back if and when our physical well-being is threatened—if our family, our community, or our country were attacked. But we shouldn't hurt other people unless it is in self-defense or in the defense of another against unchecked aggression.

Libertarian philosophers call this the Non-Aggression Principle (NAP). Don't start a fight, but always be prepared, if absolutely necessary, to finish a fight unjustly instigated by someone else. Here's how Murray Rothbard put it:

> *The fundamental axiom of libertarian theory is that no one may threaten or commit violence ("aggress") against another man's person or property. Violence may be employed only against the man who commits such violence; that is, only defensively against the aggressive violence of another. In short, no violence may be employed against a non-aggressor. Here is the fundamental rule from which can be deduced the entire corpus of libertarian theory.*[2]

Justice, says Adam Smith, is based on a fundamental respect for individual life. "Death is the greatest evil which one man can inflict upon another, and excites the highest degree of resentment in those who are immediately connected with the slain," he writes. "Murder, therefore, is the most atrocious of all crimes which affect individuals only, in the sight both of mankind, and of the person who has committed it." [3]

We all agree that the first legitimate role of government force is to protect the lives of individual citizens. But things get more complicated when it comes to defending against "enemies foreign and domestic."

In his 1796 Farewell Address, George Washington warned Americans not to "entangle our peace and prosperity in the toils" of foreign ambitions, interests, and rivalries. "It is our true policy to steer clear of permanent alliances with any portion of the foreign world."

Our first president was hardly an isolationist, and his foreign policy views were guided, in large part, by common sense and pragmatism. One of his key considerations was the budgetary implications of overly ambitious foreign entanglements. "As a very important source of strength and security, cherish public credit," Washington counseled. "One method of preserving it is to use it as sparingly as possible, avoiding occasions of expense by cultivating peace."

You might interpret Washington's skepticism, in a modern context, as warning against open-ended nation-building quagmires. Can we really establish a constitutional democracy in Iraq? Can we successfully mediate the violent disputes of warring factions in civil wars like the one going on today in Syria? Better yet, should we?

The principle of nonaggression means that we should only declare war on nations demonstrably seeking to do us harm. The men and women who volunteer for our military should not be put in harm's way by their commander-in-chief without a clear and just purpose, without a plan or without an endgame. This is just common sense.

In an era in which our enemies are no longer just confined to nations, the other key question is the balance between security at home and the protection of our civil liberties, particularly our right to privacy and our right to due process. Massive expansions of the government's surveillance authorities under the Patriot Act and recent amendments to the Foreign Intelligence Surveillance Act have civil libertarians of all ideological stripes worried that the government has crossed essential constitutional lines.

Defending America against the unchecked aggression of our enemies is a first responsibility of the federal government, but respecting the rights of individual citizens and checking the power of unelected employees at the National Security Agency is an equally important responsibility. I stand with Ben Franklin on this question. He said: "Those who would give up essential liberty to purchase a little temporary safety deserve neither liberty nor safety."

We should always be skeptical of too much concentrated power in the hands of government agents. They will naturally abuse it. Outside government, an unnatural concentration of power—such as the extraordinary leverage wielded by mega-investment banks or government employees unions—is always in partnership with government power monopolists.

2. DON'T TAKE PEOPLE'S STUFF

Life. Liberty. Property. While most of us are totally down with the first two tenets of America's original business plan, the basis of property rights and our individual right to the fruits of our labors seems to be increasingly controversial. Do we have a right to our own stuff?

In our personal lives, taking from one person, by force, to give to another person is considered stealing. Stealing is wrong. It's just not cool to take other people's stuff, and we all agree that ripping off your neighbor, or your neighbor's credit information online, or your neighbor's local bank, is a crime that should be punished.

"There can be no proper motive for hurting our neighbour, there can be no incitement to do evil to another, which mankind will go along with, except just indignation for evil which that other has done to us," argues Adam Smith. "To disturb his happiness merely because it stands in the way of our own, to take from him what is of real use to him merely because it may be of equal or of more use to us, or to indulge, in this manner, at the expense of other people, the natural preference which every man has for his own happiness above that of other people, is what no impartial spectator can go along with."[4]

But what if the stealer in question is the federal government? Is thieving wrong unless the thief is our duly elected representation in Washington, D.C., or some faceless "public servant" working at some alphabet-soup agency in the federal complex?

It seems to me that stealing is always wrong, and that you can't outsource stealing to a third party, like a congressman, and expect to feel any better about your actions.

In the real world, where absolute power corrupts absolutely, there are no good government thieves or bad government thieves. There is only limited or unlimited government thievery.

The alternative to outsourced government thievery is a world where property rights are sacrosanct, where the promises you make to others through contracts are strictly enforced, and where the rule of law is simple and transparent and treats everyone the same under the laws of the land.

Government is, by definition, a monopoly on force.[5] Governments often hurt people and take their stuff. That's why the political philosophy of liberty is focused on the rule of law. Government is dangerous, left unchecked. Consider the way too many examples from modern history to see the murderous results of too much unchecked government power: communists, fascists, Nazis, radical Islamist theocracies, and a broad array of Third World dictators who hide behind ideology or religion to justify the oppression and murder of their countrymen as a means to retain power.

All of these "isms" are really just about the dominance of government insiders over individuals, and the arbitrary rule of man over men. Unlimited governments always hurt people and always take their stuff, often in horrific and absolutely unintended ways. The architects of America's business plan were keenly aware of the dangers of too much government and the arbitrary rule of man. James Madison states it well in *Federalist* 51:

> *But what is government itself, but the greatest of all reflections on human nature? If men were angels, no government would be necessary. If angels were to govern men, neither external nor internal controls on government would be necessary.*

Government should be limited, and it should never choose sides based on the color of your skin, who your parents are, how much money you make, or what you do for a living. And it should never, ever choose favorites, because those favorites will inevitably be the vested, the powerful, and the ones who know somebody in Washington, D.C.

That's why our system is designed to protect individual liberty. "[I]n the federal republic of the United States," Madison writes, "all authority in it will be derived from and dependent on the society, the society itself will be broken into so many parts, interests, and classes of citizens, that *the rights of individuals,* or of the minority, will be in little danger from interested combinations of the majority. In a free government the security for civil rights must be the same as that for religious rights."

3. TAKE RESPONSIBILITY

Should you wait around for someone else to solve a problem, or should you get it done yourself? Liberty is an individual responsibility. The burden always sits upon your shoulders first. It is that inescapable accountability that stares you in the mirror every morning. If it didn't get done, sometimes there's no one to blame but yourself.

Free people step up to help our neighbors when bad things happen; no one needs to tell us to do that. We defend, sometimes at great personal sacrifice, what makes America so special. Freedom works to make our communities a better place, by working together voluntarily, solving problems from the bottom up.

This is the "I" in community. Communities are made up of individuals and families and volunteers and local organizations and time-tested institutions that have been around since long before you were born. All of these things work together to solve problems, build things, and create better opportunities. But notice a pattern that should be self-evident: Families are made up of free people. So are churches and synagogues, local firehouses and volunteer soup kitchens, and the countless community service projects that happen every weekend. All of these social units, no matter how you parse it, are made up of individuals working together, by choice. It does take a village, but villages are made up of people choosing to voluntarily associate with one another.

I was introduced to the philosophy of liberty by Ayn Rand. I found her work compelling because it focused on individual responsibility. Do you own yourself and the product of your work, she asked, or does someone else have a first claim on your life? I thought the answer was obvious.

Rand's critics love to attack her views that individuals matter, and that you have both ownership of and a responsibility for your own life. They usually set up a straw man: the caricature of "rugged individualism" and the false claim that everyone is an island, uncaring of anyone or anything, willing to do anything to get ahead.

"Ayn Rand is one of those things that a lot of us, when we were 17 or 18 and feeling misunderstood, we'd pick up," Barack Obama tells *Rolling Stone*. "Then, as we get older, we realize that a world in which we're only thinking about ourselves and not thinking about anybody else, in which we're considering the entire project of developing ourselves as more important than our relationships to other people and making sure that everybody else has opportunity—that that's a pretty narrow vision. It's not one that, I think, describes what's best in America."[6]

Of course it isn't, Mr. President. In Obama's simplistic configuration, there is only the "narrow vision" of the individual, and the seemingly limitless wisdom of the collective. Progressives and advocates of more government involvement like to suggest that there is a dichotomy, or at least a direct trade-off, between individual liberty and a robust sense of community.

It's easy to kick down straw men, I suppose, but the real question stands: Can governments require that people care, or force people to volunteer? It seems like such a silly question, but some seem to think the answer is "yes."

Some people just don't see the link between individual initiative and the cohesion of a community.

Justice means treating everyone just like everyone else under the laws of the land. No exceptions, no favors. "Social justice," as best I can tell, means exactly the opposite. It means treating everyone differently, usually by redistributing wealth and outcomes in society by force.

The term "social justice" was first coined by the Jesuit philosopher Luigi Taparelli d'Azeglio, who argued, "A society cannot exist without an authority that creates harmony in it." Someone needs to be in charge, he assumed, and someone needs to direct things. President Franklin Delano Roosevelt quoted Taparelli in a speech in 1932, to help justify the extraordinary, and often unconstitutional, actions taken by his administration to consolidate power in the federal government: "[T]he right ordering of economic life cannot be left to a free competition of forces. For from this source, as from a poisoned spring, have originated and spread all the errors of individualist economic teaching."[7]

Forty years later, John Rawls would expand on this idea in his influential book *A Theory of Justice*. "Social and economic inequalities," he asserted, "are to be arranged so that they are to be of the greatest benefit of the least-advantaged members of society."[8]

Can you mandate compassion? Can you outsource charity by insisting that the political process expropriate the wealth of someone you don't know to solve someone else's need? Austrian economist F. A. Hayek, ever quick to spot the logical flaws of his ideological opponents, said that social justice was "much the worst use of the word 'social'" and that it "wholly destroys" the meaning of the word it qualifies.[9]

The process of getting to the "right" outcomes, the properly reengineered social order, is never well defined. But the social justice crowd is convinced that some people just know better. They are certain that some people are better trusted with the power to rearrange things. As former U.S. representative Barney Frank used to say: "Government is what we call those things we do together."[10]

If you don't believe in individual liberty, things get complicated quick. "Social justice," the seeming opposite of plain old justice, requires someone to rearrange things by force. It's all about power, and who gets to assert their power over you. The rules are always situational, and your situation is always less important than the situations the deciders find themselves in. Someone else, defined by someone else's values, gets to decide.

Of course, if someone else is in charge, we always, conveniently, have someone else to blame. Not left free, we might just wait around for someone else to take care of it. We might not step up. We might not get involved. We might outsource personal responsibility to a third party, paid for with someone else's hard work and property.

Without liberty, any sense of community that binds us might just unravel.

4. WORK FOR IT

Liberty is a weight.

If you have ever tried to do something you've never done before, or tried to start a new business venture, or created new jobs and hired new workers, you know exactly what I'm talking about. The weight. The same is true for people who step up to solve a community problem or serve other folks in trouble. How about peacefully petitioning your government for a "redress of grievances," a right guaranteed by the First Amendment, only to be met by federal park police with pre-printed "shutdown" signs and plasticuffs?

These are all acts of risk taking, an attempt to serve a need or disrupt the status quo. These are acts of entrepreneurship. And it's all hard work.

But work is cool, too, and even some Hollywood superstars seem to get it. "I believe that opportunity looks a lot like hard work," Ashton Kutcher told the audience of screaming teenagers at the 2013 Teen Choice Awards in Hollywood. "I've never had a job in my life that I was better than. I was always just lucky to have a job. And every job I had was a stepping-stone to my next job, and I never quit my job until I had my next job. And so opportunities look a lot like work." [11]

Have you ever had to work for something, pushing against the disinterest and apathy of everyone around you? Maybe you were laughed at, but it didn't really matter. You were out to prove yourself right. To create something. To achieve something. Entrepreneurs often fail, take their lumps, and move forward to disrupt the status quo. We don't know what we don't know, but entrepreneurs have the extraordinary judgment to see around the next corner.

"What distinguishes the successful entrepreneur and promoter from other people is precisely the fact that he does not let himself be guided by what was and is, but arranges his affairs on the ground of his opinion about the future," says the great free market economist Ludwig von Mises. The entrepreneur "sees the past and the present as other people do; but he judges the future in a different way. . . . No dullness and clumsiness on the part of the masses can stop the pioneers of improvement. There is no need for them to win the approval of inert people beforehand. They are free to embark upon their projects even if everyone else laughs at them." [12]

Entrepreneurship can be a lonely business. It's hard work. Entrepreneurship is knowing that a particular problem won't be solved unless you solve it.

Part of being an entrepreneur is ignoring the naysayers, and staying fixed on a singular goal, looking around the corner of history and envisioning a better future. Working for it means responding to customer demand or creating solutions to still-unknown demands, seeing something that others can't see but still wondering if you will fail.

Do you think our founding entrepreneurs were anxious when they put their "John Hancocks" on that parchment? They pledged their lives, fortunes, and sacred honor for a principle—that people should be free—utterly ignoring their slim odds of success.

It's not so easy creating jobs, hiring new workers that become your extended family, and then lying awake at night wondering if you will make payroll on Friday. But that's what working for it is all about.

Work is hard.

But the upside of work is so awesome. It's all about the infinite potential that sits right around the next corner. You can go get it. You are free to work in pursuit of your own happiness, to associate with whomever you like, to take care of loved ones as your first priority, and to join in voluntary association with your neighbors, or your countrymen, in common cause, to make things better. Or not. It is up to you.

For all of the debate about "the rich" paying their fair share, the real question we are arguing about in America is not about the proper redistribution of the diminishing spoils between rich and poor. Every country throughout history has had its privileged class, usually favored and protected by government cronies. The real question is more fundamental: Are we still a country where anyone can get rich, where there are no government-enforced class distinctions that prevent the poor from climbing the economic ladder?

Jonathan Haidt, a professor of psychology at the University of Virginia, suggests that there is a good dose of karma in a book I coauthored in 2010, *Give Us Liberty*. "It is the Sanskrit word for 'deed' or 'action,' and the law of karma says that for every action, there is an equal and morally commensurate reaction," he writes in the *Wall Street Journal*.[13] "Kindness, honesty and hard work will (eventually) bring good fortune; cruelty, deceit and laziness will (eventually) bring suffering." My opposition to Wall Street bailouts for the irresponsible and politically gamed rules that punish hard work? "Capitalist karma, in a nutshell," Haidt concludes.

CALL IT WHATEVER YOU like. Liberty defends "the minority," the opportunity to work for it, the "underclass" with absolutely no political pull, the unconnected, and the rights of every single individual to make it. Liberty is color-blind.

Liberty is a merit-based system, and it blindly measures all of us based on the content of our character.

Why would anyone want to live life any other way but free?

5. MIND YOUR OWN BUSINESS

Free people live and let live. Free people don't have any great designs on the freedoms of other people, and we expect them to return the favor. I figure I have enough on my plate just keeping myself straight, protecting the people I love, getting my work done.

How I live my own life, and how I choose to treat others, matters. How I achieve my goals defines who I am and who I will be on the day I die. As best I can, the hows and whats in my life hopefully reflect my core principles.

But is it really any of my business to mind the business of the millions of other people working out their own dreams? I don't think so. I don't have to accept their choices or their values. But as long as they tolerate mine, as long as they don't try to hurt me or take my stuff, or try to petition the government to do it for them, why should I care?

Certainly other people will disagree with my live-and-let-live attitude. But the real question is about the proper role of government in limiting my personal decisions, or dictating my values, or the practice of my religion, or the re-definition of cherished social institutions, which have been developed and defended by people coming together in common cause.

Society should never be absorbed or distorted by the state, argues Ben Rogge, the late, great libertarian professor at Wabash College. "Society, with its full network of restraints on individual conduct, based on custom, tradition, religion, personal morality, a sense of style, and with all of its indeed powerful sanctions, is what makes the civilized life possible and meaningful." Still, he argues, we do "not wish to see these influences on individual behavior institutionalized in the hands of the state. As I read history, I see that everywhere the generally accepted social processes have been made into law, civilization has ceased to advance."

> *I, Ben Rogge, do not use marijuana nor do I approve of its use, but I am afraid that if I support laws against its use, some fool will insist as well on denying me my noble and useful gin and tonic. I believe that the typical Episcopal Church is somewhat higher on the scale of civilization than the snake-handling cults of West Virginia. Frankly I wouldn't touch even a consecrated reptile with a ten-foot pole, or even a nine-iron, but as far as the Anglican Church is concerned, I am still an anti-anti-disestablishmentarian, if you know what I mean.*[14]

Can the political process better arbitrate the definition of time-tested social mores? It seems like a ridiculous question to ask about 535 men and women who can't even balance the federal budget. Why would we hope that they weigh in on the things that really matter to us personally?

I remember when the George W. Bush administration implemented its faith-based initiative as part of a campaign of "compassionate conservatism." Whatever its good intentions, this program effectively began the process of politicizing faith-based community service. It was no longer about individuals volunteering their time and money to solve problems. By 2008, this federal program became a competitive scrum for federal grants to well-connected "faith-based" organizations. Under Barack Obama, the program was renamed and repopulated with interests and organizations to better promote his administration's priorities.

Wouldn't it be better not to set up a new program that will inevitably become politicized, corrupting everything it touches?

Consider the definition of marriage. Why does the federal government have an opinion about my marriage? Why do government bureaucrats and politicians have a right to have an opinion about, or control over, the most important personal relationship in my life? Why would we want the federal government, with all of its competing agendas and interests other than your own, involved? I think it's a really bad idea, and the fact that I had to get a license to get married to the love of my life felt somehow degrading to my most sacred bond.

I was young and idealistic when Terry and I got engaged. At the time I had made my carefully researched, impeccably principled arguments about not demeaning the sacred bond between us, and how getting the government's approval was wrong. I lost, of course. We got the government's license, on the government's terms. And we got married. Let's just say that I respect my wife's authority and her grandma's authority over my life far more than I resent the federal government's claimed but illegitimate right to dictate the terms of my personal relationships.

So yes, even I compromise on principle.

Do to others what you would have them do to you. This, of course, is the Golden Rule, and you can find iterations of it throughout the New Testament of the Bible. I would like other people, and the government, to stay out of my personal business. I plan to return the favor.

6. FIGHT THE POWER

Lord Acton, the great classical liberal political philosopher, famously warned that "power tends to corrupt" and "absolute power corrupts absolutely." [15] "The chief evil is unlimited government," argues F. A. Hayek, "and nobody is qualified to wield unlimited power." [16]

This too seems like common sense, and Americans have a healthy distrust of big, obtrusive government that seems genetically encoded in our DNA. Our system of constitutional checks and balances, and adversarial and separate branches of government, is intended to limit monopoly government power.

Notice that the goal is not electing better angels to benevolently wield power *for the right reasons*. There is some confusion about this, a difference that Hayek addresses eloquently in his most important essay on political philosophy, "Why I Am Not a Conservative":

*[T]he conservative does not object to coercion or ar-
bitrary power so long as it is used for what he regards
as the right purposes. He believes that if government
is in the hands of decent men, it ought not to be too
much restricted by rigid rules. Since he is essentially
opportunist and lacks principles, his main hope must
be that the wise and the good will rule—not merely by
example, as we all must wish, but by authority given
to them and enforced by them. Like the socialist, he is
less concerned with the problem of how the powers of
government should be limited than with that of who
wields them; and, like the socialist, he regards himself
as entitled to force the value he holds on other people.*[17]

Remember that, in the European context, "liberal"
means pro-freedom. "Conservative" means something more
like what we would call *progressive.*

So there are rules. But the architects of this model always understood that accountability rested in the hands of
the customers: American shareholders who have a right, and
an obligation, to check the bad management decisions made
in Washington, D.C. Our representatives work for us, and
we should have the right to review their job performance
and fire underperformers.

The challenge of knowing what it is that our public officials are up to has always been the biggest barrier to accountability. Quite often, busy people with jobs and families and
all sorts of personal dreams and pursuits just couldn't get
good, timely information about what our representation—

our employees—were up to behind the cloistered halls of the marble Senate office buildings and windowless federal agencies. What were they doing in there? We would usually find out about bad decisions, made for the benefit of someone else's parochial interests, after the legislation was signed, sealed, and delivered.

So normal Americans were too busy, and the barriers of entry into our participatory republic were too high for us to know. But the insiders, and the well-heeled interests that wanted a special deal, or a subsidy, or a carve-out, or an earmark, or an exemption, always showed up in Washington, hat in hand. Why? Because the return on the investment made cozying up to Washington a very profitable "business" proposition. Public choice economists refer to this perverse incentive structure as the "concentrated benefits" of D.C. power players versus "dispersed costs" incurred by anyone paying taxes.

In other words, you get screwed. This isn't a Republican versus Democrat thing. It's more about who manages to get a seat at the table first. Typically, you won't find your chair available when things really matter.

This process, more than anything else, explains all of the bailouts and debt and seemingly mindless expansion of government into our personal and economic lives.

The answer, today, is to fight the power. Government goes to those who show up. The old dismal calculus of big government is being undermined by the Internet, the decentralization of knowledge, the breakup of the old media cartel, social media that lets us easily connect with other concerned and newly activated citizen shareholders. The democratization of politics is shifting power away from insiders, back to the shareholders.

But you still have to step up and take personal responsibility. No one's going to do it for you. You can't proxy-vote your shares in America's future to some third party. If you don't like the direction your country is taking, if you don't like the dominance of D.C. insiders, senators-for-life, and super-lobbyists who get special access to the West Wing, it's time to take a look in the mirror.

The burden of individual responsibility means that sometimes there's no one else to blame but yourself.

Before you convince yourself that it's impossible to change things, think about Samuel Adams, or Mahatma Gandhi, or Lech Walesa or any other lonely activist that has done the undoable through peaceful resistance to government power.

Before you tell yourself, after years of fighting, that it's just too hard, think about the price Dr. Martin Luther King Jr. paid for his willingness to step up.

This burden, the weight of liberty, is what has driven a small minority, those special few freedom fighters over history, to buck the status quo, often at extraordinary personal costs. Those who step up, in an act of lonely entrepreneurship, and fix "unfixable" problems even as the anointed experts "laugh at them." Would you be willing to risk your life, your fortune, and your sacred honor for the principle that individuals should be left free, provided that they don't hurt people and don't take their stuff?

CHAPTER 2

YOU CAN'T HAVE FREEDOM FOR FREE

IN 1977, I BOUGHT my first Rush album. I was thirteen. The title of the disc was *2112,* and the foldout jacket had a very cool and ominous red star on the cover. As soon as I got it home from the store, I carefully placed that vinyl record onto the felt-padded turntable of my parents' old Motorola console stereo. The moment I dropped the stylus, and that needle caught the groove, I became obsessed with Rush. I got obsessed with Rush like only thirteen-year-old boys can get obsessed. I turned up the volume as loud as I thought I could get away with, and I rocked.

Mom was not nearly as pleased as I was with my new discovery. I know it sounds cliché, but she was surely the most patient woman in the world. Barbara Kibbe's youngest son was what the best peer-reviewed academic journals on parenting refer to as "a handful."

My highly anticipated jam session didn't last very long that day. Mom shut it down.

So I turned down the stereo, sat down, and began to read the liner notes inside the album's cover jacket. One of the things lost in today's era of digital downloads is the ritual of reading the lyrics and the commentary that used to be an essential part of what you were buying when you purchased new music. The notes gave context and understanding to the music and helped you connect with the musicians who created the songs you listened to.

"With acknowledgement to the genius of Ayn Rand," read the text inside the cover of *2112*. What an odd name, I thought. Who is Ayn Rand?

"2112" is a song cycle that tells the story of a futuristic, tyrannical society where individual choice and initiative have been replaced by the top-down control of an autocratic regime, where all decisions are guided by "the benevolent wisdom" of the Priests of the Temples of Syrinx. The Priests boast that they've "taken care of everything" using the awesome power of their "great computers" to bestow equality on all mankind. They lord over a "nice, contented world."

In the plot of this dystopian tale, one of the "common sons" approaches his controllers with a new discovery: a guitar, an instrument that could change things for the better by providing inspiration and music. Could this "strange device" be a vehicle for individual expression? He naïvely thinks that his controllers will care, will be open to new beauty, new innovation, and more creative freedom. "There's something

here that's as strong as life," he tells them. "I know th
will reach you." Instead of hearing him out, the Priests
crush his newly found instrument under their feet, crushing
his spirit in the process. "Forget about your silly whim," the
troublemaker is told. "It doesn't fit the plan!"

In the 1970s it was virtually impossible to find out about
new music and different genres that didn't fit the one-size-
fits-all mold of commercial pop. Everything on the radio
was Top 40, predetermined to be what you wanted to hear
by some nameless, gray-suited music executive. Everything
was very top-down, and choices and information typically
flowed just one way, leaving alternatives undiscovered, un-
heard by consumers, crushed by the silence of ignorance.
You just didn't know what you didn't know. So the experts
chose for you, and in 1977 they had selected, for me, really
awesome songs like Andy Gibb's "I Just Want to Be Your
Everything," Barbra Streisand's "A Star Is Born," and Captain & Tennille's "Muskrat Love." The insipid disco version
of the *Star Wars* cantina bar song, by Meco, sat on the top of
the *Billboard* charts for two weeks, subjecting me and otherwise discerning people to its cruel torture on an endless
rotation.

Until I found Rush, that is. I actually discovered the
band as I was walking past the recreation center at my high
school. Some cool kid was playing *All the World's a Stage,* a
live album by Rush released soon after *2112*. Of course, the
record store didn't have that album when I finally convinced
my mom to drive me there, so I settled for the one with the
cool, ominous red star, the only Rush album in stock. There
were very few choices in the days of bricks and mortar—no

Internet that gave people the freedom to
ey wanted, when they wanted—and vital
should have held my much-wanted record
up Andy Gibb, Barbra Streisand, and Cap-
tain & ... le. I don't honestly remember, but I have little
doubt that there were stacks and stacks of *Star Wars and
Other Galactic Funk* by Meco.

All the World's a Stage by Rush? It didn't fit the plan.

It was as if these faceless record executives entrenched in
the Music Industrial Complex were goading me to revolution. Why did music have to suck so bad? Why did everything have to sound the same?

As it turns out, I was hardly alone in feeling this way. In
the mid-1970s, several years before I would discover *2112,*
the members of Rush were battling their own record label for
control of their artistic direction. What kind of music would
the band make? Would anyone buy it? The band wanted to
pursue its own creative path, even if it didn't fit with someone else's conception of "good" music. Mercury Records
wanted something more "commercial." They wanted Rush
to sell more records, or else. "There was a great deal of pressure on the band at that time," says Alex Lifeson, the band's
guitarist.

If you follow any genre of music, how it evolves and
mutates, you have already heard this story a thousand times.
It is the clash between tradition and innovation, and the
creative destruction that drives individuals to challenge the
status quo. Record-label executives always get squirrelly
when some difficult-to-manage artist creates new music that
deviates from the norm. Even Miles Davis, the great jazz

trumpet innovator of the 1950s, eventually would disavow the new creative directions his most important collaborator, saxophonist John Coltrane, took jazz in the 1960s. Perhaps challenged by his protégé, Davis himself redefined the genre again in the late 1960s, after Coltrane had passed at a tragically young age. Jazz critics would later attack Davis for his groundbreaking masterpiece *Bitches Brew,* released in 1970, as "commercial crap that was beginning to choke and bastardize" jazz standards.[1]

The inherent discomfort the established conventional wisdom has with musical innovation is captured perfectly, and hysterically, in the 1984 movie *Amadeus,* when Holy Roman Emperor Joseph II tells a young Mozart that the "Non più andrai" march from his 1786 opera *The Marriage of Figaro* has "too many notes."

"Cut a few," Joseph advises, "and it will be perfect."

Incredulous, Mozart asks: "Which few did you have in mind, Majesty?"

Too far. Too individualistic. Too extreme. Too many notes. You just know it's going to happen, the labels and the name-calling, the defensiveness, when the protectors of the status quo feel threatened by change and principled disruption.

When it comes to innovation, sometimes the customer is always right. But other times an innovator shakes up market perceptions and upsells buyers on a better product— a new idea that you didn't even know you needed until someone else figured it out for you. This process of creative disruption—standing on the shoulders of your intellectual forefathers all the while challenging them and their best work—seems to be where the good stuff in life comes from. And it can only happen if people are free. Free to succeed. Free to fail. Free to speak their minds and disagree with the experts. Free to choose. Think about the horseless carriage, handheld computers, or the MP3 files on your iPod that replaced CDs, that replaced cassettes and eight-track tapes, and yes, that even replaced vinyl.

This disruption seems particularly true in music. Music and freedom just seem to go together, just like the word "bacon" belongs in any sentence that includes the phrase "proper meal." I can't prove it, but you just know that it's true.

Back in 1977, such profound insights eluded me. I was still wearing black concert tees and wondering who the heck Ayn Rand was, when I stumbled upon a used copy of her novella *Anthem* at a neighborhood garage sale. I took it home and read it without putting it down once. What an awesome book it was, about a dystopian society where the word "I" had been erased by an oppressive, collective "We."

> *It is a sin to write this. It is a sin to think words no others think and to put them down upon a paper no others are to see. It is base and evil. It is as if we were speaking alone to no ears but our own. And we know well that there is no transgression blacker than to do or think alone. . . .*
>
> *Our name is Equality 7-2521, as it is written on the iron bracelet which all men wear on their left wrists with their names upon it. We are twenty-one years old. We are six feet tall, and this is a burden,*

for there are not many men who are six feet tall. Ever have the Teachers and the Leaders pointed to us and frowned and said: "There is evil in your bones, Equality 7-2521, for your body has grown beyond the bodies of your brothers." But we cannot change our bones nor our body.

We were born with a curse. It has always driven us to thoughts which are forbidden. It has always given us wishes which men may not wish.

Despite insurmountable odds, the good guys, the "cursed" ones, the ones who begin to start their sentences with the word "I," persevere. I connected with the struggle to be free—different, independent, responsible for my own successes and failures.

I immediately set out to find *The Fountainhead*, which was listed in the front pages of my dated, dog-eared paperback copy of *Anthem* as one of the "other novels" by Rand. No mention of *Atlas Shrugged*, which hadn't even been conceived of when my now-cherished copy of *Anthem* went to press. Imagine how long it took me to find a copy of *The Fountainhead*. Back in the day, you couldn't just log into your account on Amazon.com and find it, or the multitude of other books related to it. I looked in any bookstore, at every opportunity. It was difficult to find. But I was obsessed.

Neil Peart, the drummer and lyricist for Rush, was also obsessed with Ayn Rand at the time of his band's career-defining struggle with their record label. He started off reading *The Fountainhead* because "all the smart kids used to carry that around" in high school.[2] Peart "introduced her writing to us," says lead singer and bass guitarist Geddy Lee. "We all liked the book *Anthem*, which is the thing that kind of inspired *2112*."

The band had toured relentlessly in support of their last album, *Caress of Steel*, but the record had been trashed by music critics (a trend that would go on for decades).

Without the music industry press on Rush's side, album sales were disappointing. For the next album, company headquarters wanted something conventional, something that would sell. "I felt this great sense of injustice that this mass was coming down on us and telling us to compromise, and compromise was the word I couldn't deal with," recalls Peart. "I grew up a child of the 60s and I was a strong individualist and believed in the sanctity of: 'you should be able to do what you want to do, you know, without hurting anyone.' " Artistic integrity, for Peart and his bandmates, had crashed headlong into the expediency of the moment.

Instead of following the rules, instead of recording an album that conformed to the expected, Rush made *2112*. At a time when successful pop songs ran about three minutes long, a twenty-minute song cycle about totalitarian oppression on a far-away planet was hardly what that sales team at Mercury Records had in mind. "We got angry and thought, okay, if this is our last shot we are going to give it everything and we're gonna do it our way," recalls Peart. So Rush did it their way, giving it everything they had in them.

After discovering *Anthem* and *The Fountainhead*, by the time I turned fifteen I had read all of Rand's fiction and many of her nonfiction works, such as her anthology, *Capitalism: The Unknown Ideal,* in which she recommends the works of the Austrian economist Ludwig von Mises.[3] I somehow found a copy of *Human Action,* Mises's comprehensive treatise on economics, and began to read it. I didn't really know what I was doing, and maybe I understood a fraction of what I was reading, but don't ever try to tell a teenager what he can't do. I was, after all, obsessed.

As you might imagine, wearing black AC/DC concert tees, listening to Rush and Led Zeppelin and the Stones, and quoting Ayn Rand and Ludwig von Mises to anyone who would listen turned out to be the worst possible strategy for meeting girls in high school. Thanks to the stagflation of the Carter presidency and the minimum wage, I could not find a job in Grove City, Pennsylvania, when I turned sixteen. My schedule was clear! My lack of social skills, a job—and dates—provided plenty of time to read things normal kids don't.

I graduated from high school not knowing what I wanted to do. I wasn't particularly interested in going to college, but at my father's insistence I applied to a number of schools. Sumner Kibbe was obsessed—obsessiveness being an apparently hereditary trait—and I didn't typically get away with saying "no." I ultimately chose Grove City College for one simple reason: It was the cheapest. I was able to pay my tuition by clearing trees and washing dishes for the college (students were exempt from the minimum wage that had been such a barrier to my earlier entry into the workforce). I set out as a biology major, but I was bored with it. I was barely scraping by with my classwork. I was now reading Adam Smith and other "classical liberal" philosophers that I had discovered reading Mises, and that was far more interesting. I never imagined that I could pursue a degree (let alone a career) consistent with the ideas I was learning about outside the classroom. I just didn't know there were others who thought like I did, had read what I was reading.

It seems so ridiculous to admit today, but as an incoming freshman at Grove City College, I was utterly unaware of the fact that the head of the economics department, Dr. Hans Sennholz, was only one of a handful of economists who had earned his Ph.D. from Ludwig von Mises. *Human Action* was the required text for Econ 301. I walked past that department every day on my way to the science classes I was not really interested in, but I just didn't know. Talk about a "knowledge problem."

I didn't figure any of this out until a late-night argument with a friend, Peter Boettke.[4] We were in the same fraternity, and we were of course debating just how limited "limited government" should be. I know what you're thinking. Nerd. Really big nerd. Wikipedia defines a nerd as "a person, typically described as being overly intellectual, obsessive, or socially impaired. They may spend inordinate amounts of time on unpopular, obscure, or non-mainstream activities, which are generally either highly technical or relating to topics of fiction or fantasy, to the exclusion of more mainstream activities." There were few girls at the ADEL house that night, but at least they were spared an intense discussion on the proper role of government in a constitutional republic. As our argument wound down, Pete suddenly stopped to ask me, "Why aren't you an economics major?"

I didn't know.

It's remarkable how my life changed that night. I switched to economics and philosophy, and my grades immediately went from C's and D's to A's and B's. (My wife, Terry, whom I started dating around the same time, was given full credit for the miraculous turnaround in my academic performance by my parents. She was an engineer, like Pops, so she was "smart." She never disabused them of this belief. Like I said, she's a smart girl.)

My veil of ignorance was lifted, and I was quickly exposed to a body of ideas and community of people united by the values of individual freedom and the limitless potential of people when offered a chance to strive, seek, and achieve. It seemed like there were dozens, maybe hundreds of people who were thinking about liberty, individualism, and the power of ideas, just like me. Dr. Sennholz, who by that time had developed a close mentoring relationship with Dr. Ron Paul, a newish congressman representing the 14th District of Texas, also became my intellectual mentor. He introduced me to the Foundation for Economic Education, in Irvington-on-Hudson, New York, and the Institute for Humane Studies and eventually the Center for the Study of Market Processes, both at George Mason University.

I went to GMU for graduate studies in economics, again at Pete Boettke's urging. In 1984, Citizens for a Sound Economy was founded out of the Austrian economics program at George Mason, and Dr. Paul became the founding chairman. As a graduate student at Mason, I was loading trucks at UPS to pay the tab. I took a 50 percent pay cut to join CSE in 1986, but I was thrilled. I was going to get paid to fight for freedom. How cool was that?

I went on to other things, but came back to CSE in 1996. CSE became FreedomWorks on July 22, 2004. I became president of FreedomWorks that day.

Back in 1976, Neil Peart, the drummer and lyricist for Rush, was thinking about his future and pursuing his dreams. He penned the dystopian lyrics to "2112" thinking about *his* individual freedom. "I did not think of politics and I did not think of global oppression," he recalls. No, he was thinking: "These people are messing with me!" He and the rest of the band found their inspiration in *Anthem,* the same novella that had turned me on.

"You can say what you want about Ayn Rand and all the other implications of her work, but her artistic manifesto, for lack of a better term, was the one that struck home with us," says Geddy Lee. "It's about creative freedom. It's about believing in yourself."

Fans agreed. Despite its not-ready-for-pop-radio format, *2112* reached number 61 on the *Billboard* pop album charts, the first time the band had cracked the Top 100. Which is the only reason I was able to find a copy in the record stacks among the multitudinous pressings of "Muskrat Love."

Creative freedom aside, the brief note inside the sleeve of *2112,* the one hat-tipping Ayn Rand, set the world of music experts—the critics—afire with ideological rage. H. L. Mencken once described a historian as "an unsuccessful novelist,"[5] referring to the propensity of some historians to make it up as they go along. Similarly, you might characterize music journalists as frustrated musicians that slake their bitterness on youth. That was certainly the case with Barry Miles, a music critic writing for England's *New Music Express*, who had a philosophical ax to grind in his trashing of Rush that had nothing to do with the quality of the music they made.

It was right out of a scene in *The Fountainhead,* where self-styled architectural critic and committed hater of intellectual achievement Ellsworth Toohey decides to destroy the young architect Howard Roark with words. On page 7 of the March 4, 1978, issue of *NME,* the headline read "Is Everyone Feeling All RIGHT (Geddit?)" As someone who reads the music press, this ranks as one of the most hateful hit pieces on a band I have ever seen. The problem, it seems, was the source of the band's ideas. Neil Peart is quoted, arguing that his band is "certainly devoted to individualism as the only concept that allows men to be happy, without somebody taking from somebody else."[6] The article gave short shrift to Rush's music. No, this was a hit piece and a clumsy vehicle for a hack journalist to express uninformed disdain for Neil Peart's developing libertarian ideology:

> *So now I understood the freedom they are talking about. Freedom for employers and those with money to do what they like and freedom for the workers to quit (and starve) or not. Work makes free. Didn't I remember that idea from somewhere? "Work Makes Free." Oh yes, it was written over the main gateway to Auschwitz Concentration Camp.*

... principles that firmly apply to every ... otes Peart again as saying. "I think a ... hat way. That you have a guiding set ... bsolutely immutable—can never be ... That's the only way."

"Shades of the 1,000 Year Reich?" observes a very bitter Miles, darkly.

"This journalist," recalls guitarist Alex Lifeson, "wrote it up like we were Nazis, ultra-right-wing maniacs."

Really? Auschwitz? Shades of the Third Reich? Nobody likes being called a Nazi—except, I suppose, Nazis. For the rest of us, it is a conversation stopper, one of the deepest insults one can hurl, like "racist." A "Nazi" is more than a "national socialist" or even a "fascist." No, a "Nazi" is a cold-blooded mass murderer.

Of course, "individualism" as described by Ayn Rand or Neil Peart or anyone else for that matter is the very antithesis of national socialism or any ideology that enables a government act of mass murder. I think the accusers who smear others with Nazism know that, and the real purpose is to stigmatize their philosophical enemies. Saul Alinsky, the radical community organizer from Chicago, said it best in *Rules for Radicals*.

Rule number 5: "Ridicule is man's most potent weapon."

Rule number 13: "Pick the target, freeze it, personalize it, and polarize it."

Well, the *New Musical Express* certainly personalized it: Both of Geddy Lee's parents had been teenage prisoners held at Auschwitz. "I once asked my mother her first thoughts upon being liberated," Lee told a reporter for

JWeekly in 2004. "She didn't believe [liberation] was possible. She didn't believe that if there was a society outside the camp how they could allow this to exist, so she believed society was done in."[7] The article goes on:

> *In fact, when Manya Rubenstein looked out the window of a camp building she was working in on April 15, 1945, and saw guards with both arms raised, she thought they were doing a double salute just to be arrogant. She did not realize British forces had overrun the camp. She and her fellow prisoners, says Lee, "were so malnourished, their brains were not functioning, and they couldn't conceive they'd be liberated."*
>
> *It is easy to see why Manya Rubenstein had given up on civilization. She and future husband Morris were still in their teens—and strangers to one another—when they were interned in a labor camp in their hometown of Staracohwice (also known as Starchvitzcha), Poland, in 1941. Prisoners there were forced to work in a lumber mill, stone quarry, and uniform and ammunition manufacturing plants.*
>
> *From Staracohwice, about an hour south of Warsaw, Manya and Morris, along with many members of both their families, were sent to Auschwitz. Eventually Morris was shipped to Dachau in southern Germany, and Manya to Bergen-Belsen in northern Germany. Thirty-five thousand people died in Bergen-Belsen from starvation, disease, brutality and overwork, according to information from the U.S. Holocaust Memorial Museum. Another 10,000 people, too ill*

and weak to save, died during the first month after liberation.

His parents' heroic struggle against Nazi genocide really defined Geddy Lee's upbringing in Toronto, and their experiences were discussed openly. "These were the things that happened to them during the most formative time in their lives," he says. "Some people go to horseback riding camp; my parents went to concentration camp."

Can you imagine his reaction to Barry Miles's ad hominem "Nazi" smears against the band in 1978? "Just so offensive," says Lee, in his typical, understated way.

Ayn Rand, like Geddy Lee, had firsthand knowledge of just how deep such smears can cut. Born Alissa Rosenbaum, Rand was growing up in St. Petersburg, Russia, when the communists took power in 1917. Her Jewish family "endured years of suffering and danger" after her father's small business was confiscated. She wanted to be a writer, but saw no hope for that under a new government regime where the freedom to express opinions, to question authority, to think for yourself, was prohibited. With the help of her family, she fled communist Russia for the United States, arriving when she was twenty-one years old.

"To free her writing from all traceable associations with her former life," observes Stephen Cox, "she invented for herself the name Ayn Rand and set out, like the hero of [*Anthem*], to make a new life for herself, in freedom."[8]

The critics never really warmed up to Rand's work, just like they never really warmed up to Rush's music. More than their art, I suspect it was their combative individualism that really irked the critics. As Gore Vidal noted in his contemptuous review of *Atlas Shrugged,* the book was "nearly perfect in its immorality." For Rand—as for Rush—there was a price to be paid for pursuing her chosen path in life. Challenging the status quo, and the freedom to do so, all came at a price. Freedom, for them, was not free. There was a downside, and it might have been easier to give in and comply with the expectations of others.

But the upside to freedom is so much better. Fans, customers hungry for something else, found Rush just like they found Rand.

The critics may have resented their work, but fans, customers hungry for something else, found them. It is said that *Atlas Shrugged,* Rand's magnum opus, is the second-most influential book in history, a distant second to the Bible.[9] According to the Recording Industry Association of America, *2112* has sold more than 3 million copies since it was released, a triple-platinum record. Overall, Rush has sold some 40 million records, and the band ranks third, behind the Beatles and the Rolling Stones, for the most consecutive gold or platinum studio albums by a rock-and-roll band.

And it all started with *2112.* It started with a willingness to stand on principle when the easier path was compromise. It started, incidentally, "with an acknowledgement to the genius of Ayn Rand." The band took off, fueled by music fans looking for something different, something inspired by disruptive innovation and creative freedom.

My personal tastes in music, like the books I was reading, eventually branched out to many different genres. I got into the Grateful Dead. If you don't get the Dead, you likely never saw the band live. There was a profound sense of community between the players onstage and their audience. Jerry Garcia, the iconic lead guitarist for the Dead, often spoke of his musical influences, including jazz, bluegrass, and blues. As a player, Garcia was very immersed in American musical traditions, and his opinions led me to Miles Davis and John Coltrane, and even bluegrass.

I particularly liked the spontaneous nature of the Dead's jams and the way Coltrane's quartet would explore the outer bounds of jazz structure. There were very few rules to guide, but plenty of room for individuality and exploration. The resulting interplay between musicians, sometimes leading, sometimes following, was a perfect metaphor for the peaceful cooperation of individuals working together towards a common goal greater than the sum of its parts. The music seemed analogous to the free association between individuals in a civil society, the interplay between institutional rules and creative disruption that Hayek and his protégés would dub the "spontaneous order." My musical interests, in a sense, tracked my expanded understanding of the ideas of freedom.

I really didn't revisit my early obsession with Rush until 2010, when an insurgent Senate candidate named Rand Paul began playing the band's "Spirit of Radio" at campaign events. He's a big fan, it turns out.

"I grew up in a libertarian family," the now well-known senator from Kentucky told me when I had a chance to sit down with him in 2013. "Ayn Rand was on a lot of different bookshelves. I read Ayn Rand when I was seventeen. I was probably a Rush fan before that, but I already knew of Ayn

Rand. So to me the serendipity was that I actually liked this band that knew about Ayn Rand. I remember reading the lyrics to *2112* and then reading *Anthem* and saying this is basically *Anthem* in music."

As it turns out, the lawyer for Rush's record label is not, apparently, a big fan of Rand Paul. Robert Farmer, general counsel for the Anthem Entertainment Group Inc. in Toronto, issued the following statement in response to the candidate's musical choices at events: "The public performance of Rush's music is not licensed for political purposes: any public venue which allows such use is in breach of its public performance license and also liable for copyright infringement."

The warning was issued after a reporter from *The Atlantic* pressed the issue.[10]

Okay, so maybe the band just doesn't like politics. Maybe they respect their fans enough not to choose sides. Maybe, as their song "Tom Sawyer" goes, "His mind is not for rent, to any god or government."

Or maybe it really sucks being called a Nazi. Maybe the hate cuts deep when it's so personal, so unfair, so offensive. Maybe they just want to do their work.

Ever since that ridiculous, slanderous, and, yes, hurtful article was published—just as their hard work as musicians was starting to pay off—it seems that the band members have had to answer the same question, over and over: "Are you guys really ultra-right-wing lunatics?"

In 2012, Neil Peart was giving a rare interview to *Rolling Stone* to talk about the band's new album, *Clockwork Angels*. He's not a talker, and typically "doesn't like all of the hoopla." But he really wanted to talk about his latest work. Of course, the question came up again. Do you *really* like Ayn Rand?

He says:

> For me, it was an affirmation that it's all right to to-
> tally believe in something and live for it and not com-
> promise. It was as simple as that. . . . Libertarianism
> as I understood it was very good and pure and we're all
> going to be successful and generous to the less fortunate
> and it was, to me, not dark or cynical. But then I soon
> saw, of course, the way that it gets twisted by the flaws
> of humanity. And that's when I evolve now into . . .
> a bleeding heart Libertarian. That'll do.[11]

That'll do. I'm a bleeding heart libertarian, OK? You can almost hear the resignation in his voice. *Can we talk about my work now?*

I found some personal inspiration in seeing Rush play live in 2013 in Austin, Texas. I hadn't seen the guys for quite some time. Work and life got in the way. They still have incredible passion and talent, and their audience is still one of the most connected as a community, with the band, in all of live rock music.

I started thinking about them again in the midst of par-ticularly challenging times for me and my extended family at FreedomWorks. The critics were calling us names. They were trying to smear us. We were "too uncompromising." We were too "pure." And that was coming from our supposed friends. We were willing to hold both Democrats and Republicans to the same standard instead of just picking sides that were artificial. We helped hold a number of politicians accountable to their shareholders, the voters. We were in the process of repopulating Washington, D.C., with more principled representation, young leaders more accountable to the principles of liberty.

Somewhere along the way, we apparently pissed off somebody really important. To this day, I'm not sure who exactly tried to take us out. But it was a hard time, and some of the personal attacks cut deep.

You see, I work in a town, Washington, D.C., that values compromise over principle. The streets that crisscross the nation's capital are lined with buildings filled with people who make a lot of money getting special favors from the political process. A typical meeting with an elected official begins with a question: "What can I do for you?" In reality, the question really being asked is "What can you do for me?" Compromise is the currency, because that's how everyone gets paid. Everyone wants something from someone. Everyone is looking for your "tell," the Achilles' heel that makes you wobbly enough, wanting the money and the power and the influence. Wanting to cut a deal. To compromise.

I remember debating Chris Matthews, the guy on MSNBC's *Hardball,* once at an event in Aspen. I was making a (surely profound) point, and Matthews abruptly interrupted. He does that. "I know, I know," he said. "I read Ayn Rand in high school. I used to believe that stuff, too, but then I grew up." Maybe he didn't know he was parroting his favorite president, Barack Obama.

I've heard this so many times. I'm sure you have, too. I suppose Neil Peart heard it more than most when he was trying to live down the youthful enthusiasm for liberty he shared with a dishonest critic in 1978. *Grow up. Play ball. Get in line.*

Well, I don't want to "grow up." I don't want to if growing up means abandoning the principle that individuals matter, that you shouldn't hurt people or take their stuff. I don't want to give up on values that have gotten me down the road of life this far. I won't "grow up," if that means not seeking ideals, taking chances, and taking responsibility for my own failures. I don't want to compromise, at least not on the things that really matter. I don't want to split the difference on someone else's bad idea, and then pat myself on the back for "getting something done."

I have no plans to fall in line.

I do the best that I can, and I belong to a community of many millions of people who seem to agree with me on the things that really matter. And we are going through this test together. Not compromising seems to be the glue that holds us as a social movement. Alone you might buckle, but are you really willing to let all of us down?

Many people in Washington, D.C., want to stop us. Sometimes they call us names, names meant to damage and hurt. Should we let them? Should we back down, or take the easier path? I can only think back to that afternoon in 1977, lying on my back on my parents' plush red wall-to-wall carpeting. *"You don't get something for nothing."* The final song on the second side of *2112* is playing. It's called, appropriately, "Something for Nothing." I'm listening, reading the lyrics inside the record sleeve, the one with the cool, ominous red star. *"You can't have freedom for free."*

CHAPTER 3

THEM VERSUS US

ON AUGUST 28, 1963, Dr. Martin Luther King Jr. delivered the speech of his life.

"I am happy to join with you today in what will go down in history as the greatest demonstration for freedom in the history of our nation,"[1] he began. MLK was, of course, addressing some 250,000 people who had joined together for the March on Washington for Jobs and Freedom. "When the architects of our republic wrote the magnificent words of the Constitution and the Declaration of Independence," he told the crowd, "they were signing a promissory note to which every American was to fall heir. This note was a promise that all men, yes, black men as well as white men, would be guaranteed the unalienable rights of life, liberty, and the pursuit of happiness."

It had been a long journey to the steps of the Lincoln

Memorial in Washington, D.C., and many black Americans had suffered, and died, along the road to that moment. But King eloquently rejected calls to meet the police dogs, fire hoses, billy clubs, and tear gas in kind. "We must not allow our creative protest to degenerate into physical violence," he implored.

King, who was the president of the Southern Christian Leadership Conference (SCLC), a nonprofit organizer of the march, eventually put aside his prepared remarks and proceeded to deliver the most eloquent call for equal treatment under the law ever spoken: "I still have a dream. It is a dream deeply rooted in the American dream," he told the crowd. "I have a dream that my four little children will one day live in a nation where they will not be judged by the color of their skin but by the content of their character."

The next day, William Sullivan, the chief of the Federal Bureau of Investigation's domestic intelligence division, penned an internal memo: "Personally, I believe in the light of King's powerful demagogic speech yesterday he stands head and shoulders over all other Negro leaders put together when it comes to influencing great masses of Negroes. We must mark him now, if we have not done so before, as the most dangerous Negro of the future in this Nation."[2]

It was the eloquence of the speech. MLK had connected with a fundamental American value, that everyone should be treated equally under the laws of the land. Because he spoke out, he became "dangerous." He was deemed a threat, so he would not be treated equally under the law by agents of the U.S. government. He would be singled out, targeted by government bureaucrats. He had to be stopped.

The FBI's obsession with MLK's rising star started at the top of the FBI, with Director J. Edgar Hoover. In a clandestine campaign against King—against an American citizen attempting to practice his First Amendment rights to petition the government for a redress of grievances, to peaceably assemble, and to speak freely—a cabal of powerful federal government bureaucrats with extraordinary discretionary power proceeded to stalk, persecute, and smear a man they viewed as an enemy to their interests. "FBI officials viewed the speech as significantly increasing King's national stature," says MLK historian David J. Garrow. After August 28, he became "measurably more 'dangerous' in the FBI's view than he'd been prior."[3]

On October 10, Hoover convinced the attorney general of the United States to authorize wiretaps on MLK's phone as well as the office phones of the SCLC. The official rationale was their suspicion that MLK was collaborating with communist sympathizers. The attorney general at the time, the top law enforcement officer in the nation, was Robert F. Kennedy, brother and close confidant to President John F. Kennedy. Wiretapping King's phone was perhaps one of RFK's most ignominious acts.

Of course, by December 1963, Hoover went well beyond what the Kennedy administration had authorized, and began installing microphones in the hotel rooms where King was staying. One conversation, taped in May 1965 and released in 2002, captured a conversation between King and Bayard Rustin regarding a dispute between the SCLC and the Student Nonviolent Coordinating Committee over a proposed statement of coalition unity.

"There are things I wanted to say renouncing communism in theory but they would not go along with it," complains King. "We wanted to say that it was an alien philosophy contrary to us but they wouldn't go along with it."

The FBI failed to disclose this information to the White House, instead using its illicit snooping to intimidate, threaten, and blackmail King.[4] Information that the FBI gathered about MLK's personal behavior was used in vicious attempts to control him, to silence him, to break up the coalition he was struggling to hold together, to stop him. By any means necessary.

FREEDOM, OR POWER?

Does it ever make sense to give so much unchecked power and authority to government agents? Can we trust them to *be better* than the rest of us? Can we trust them to *know better*?

I say no. This book argues for more individual freedom and for limiting the discretionary power of government. Too much power corrupts. Absolutely.

And J. Edgar Hoover's iniquitous behavior proves my point. The treatment of MLK certainly meets my definition of government tyranny.

I believe that there is a growing awareness among people in America, and all over the world, that governments have too much power, and that power is abused. Individual freedom, choice, upward mobility, and voluntary cooperation among free people is the better approach. In a world that is rapidly decentralizing access to information, lowering barriers to entry, barriers to knowing, freedom works even better today than it did in 1776.

Others argue the opposite, that the fear of runaway government power is outdated, that America has outgrown the old model based on liberty. It is time to reject an abiding skepticism of too much central control, they say, and let the benevolent redesigners work their magic.

They say: *More government involvement in our lives is essential to offset concentrated market power and corrupt businessmen and anyone else who might take advantage. People can't be trusted with freedom. Besides, freedom is messy and chaotic, and we won't always make the right choices. We won't always like the way things turn out, the way wealth and resources are allocated. Government can fix these problems. We just need to make sure that the power rests in the hands of the right people. There are good guys and bad guys. The right public servants can be trusted to rein in the greedy hordes.*

This was the pipe dream of "progressives" going back to the late 1800s. Well-paid civil servants, with all the right pedigrees, from all the right families, and protected from political judgment and the push and pull of democracy, would be given the power and the resources to better manage things from the top down.

The architects of this country were pretty clear on these questions. The author of that "promissory note" that Dr. King referred to in 1963 while standing on the steps of the Lincoln Memorial, Thomas Jefferson, wholly embraced the genetically ingrained American skepticism of government power and an idealistic belief in dispersing authority across society, from the bottom up. The power should be with individuals, Jefferson believed, with "We the People."

The founders were very much a product of, as well as advocates for, "the Spirit of '76." "Government is not reason," warned our first president, George Washington. Government "is not eloquent; it is force. Like fire, it is a dangerous servant and a fearful master."

That was then, says the progressive historian Joseph J. Ellis. Today, the really sophisticated thinkers, the ones with the right academic pedigrees, are shedding their fear of big government. The divide is clear, says the Mount Holyoke professor, between those who view the government as "them" and those who view government as "us." Them versus Us. On this question there is little doubt where he stands: "The expanding role of government in protecting and assuring our 'life, liberty and the pursuit of happiness' has become utterly essential."

There it is again. "Life, liberty and the pursuit of happiness." For Jefferson, this immortal phrase held up the freedom of the individual and self-determination, the opportunity to be whatever you can make of yourself. Respecting your liberty was the first duty of government, and in 1776 it was a radical concept. These were "unalienable rights," rights that we Americans were free to pursue unbound by government roadblocks. MLK invoked the phrase in 1963 to redeem "a promissory note": freedom from unequal treatment under the law, from government-imposed discrimination, and the promise of a "color-blind society."

Now the real promise is a guaranteed "right" to bigger government? Ellis was reacting to what he describes as the "libertarian" distrust bubbling up from the grass roots circa 2009. Tea party activists were expressing, in no uncertain terms, that government had gotten too big, that it was too involved in everything from big bank bailouts to redesigning our access to health-care services. In 2009, this protest movement, just like the original Boston Tea Party in 1773, seemed to be taking on a life of its own, and progressive advocates for more government oversight of your life didn't like it. Not one bit.

This was the same meme of the times coming from Democrats (and many establishment Republicans as well): There was something slightly dangerous about the new surge of liberty-mindedness emerging through the grass roots. And it wasn't just academics who were expressing concern. Right after Tax Day in 2009, Senior White House Advisor David Axelrod told CBS's *Face the Nation* that the tea party represented "an element of disaffection that can mutate into something that's unhealthy."[5]

Message received.

This is the "progressive" mind-set: *Freedom, as a governing philosophy, is just old-fashioned, past its use-by date. Anachronistic. Today, we know better. The right people, the smart, good people, can be trusted to get government right. They just need our trust, our money, and more power. Old superstitions and a libertarian skepticism of centralized power are getting in the way of progress.*

AWESOME AUTHORITY

This "shut up and trust us" narrative was picked up by Barack Obama again in a speech on May 5, 2013. His commencement address to the graduating students of Ohio State University scolds those of us who would question his grand vision:

> *Unfortunately, you've grown up hearing voices that incessantly warn of government as nothing more than some separate, sinister entity that's at the root of all our problems; some of these same voices are also doing their best to gum up the works. They'll warn that tyranny is always lurking just around the corner. You should reject these voices.*

Few dispute the president's way with words. But sometimes you have to break things down to get at their meaning. As a rule, you always know to pitch all of the words that come before the inevitable "but." Just disregard them. Erase the qualifying words from your mind to get at his point: "We have never been a people who place all of our faith in government to solve our problems; we shouldn't want to. *But* we don't think the government is the source of all our problems, either."

We don't think the government is the source of all our problems.

If government is not the problem, it must be part of the solution, right? I am reminded of the famous command from Captain James T. Kirk to the starship *Enterprise*'s chief engineer: "Scotty, I need more power."

It's all part of a better, bigger plan.

"The founders trusted us with this awesome authority," continues the most powerful man in the world in his commencement address at Ohio State. "We should trust ourselves with it, too."

Did the founders entrust us with *awesome authority*? Do we trust one man, any man, in this case a man named Barack Obama, with *awesome authority*? Should we? Would we have wanted to trust that man if his name was George W. Bush? Or Ronald Reagan?

I think the founders entrusted us with *awesome responsibility*, the responsibility of freedom, not *awesome authority* in someone else's hands. I think people should live their own lives and pursue their own happiness free from too much government meddling.

Since 2009, I have been part of a rapidly growing community of folks who agree with me that freedom works; they have been stepping out from across the ideological spectrum. They are worried that the federal government is out of control. That it is becoming all about them, not us.

And it took Lois Lerner to prove us right, and "Them" wrong. Again.

YOU ARE THE TARGET

Lerner, of course, was the Internal Revenue Service director in charge of tax-exempt organizations, who would infamously plead the Fifth during her testimony before the House Oversight Committee on May 22, 2013.

On May 10, just five days after Obama's "awesome authority" speech, Lerner dropped the bombshell admission that put her in the hot seat before Congress. Speaking at an American Bar Association conference, she used an audience question to "apologize" for the inappropriate targeting of conservative and libertarian activist groups prior to the presidential election of 2012. Innocent mistakes were made, she concedes. But it wasn't her fault. She threw "our line people in Cincinnati" under the bus for their "not so fine" targeting of tea partiers. "Instead of referring to the cases as advocacy cases, they actually used case names on this list," she said. "They used names like Tea Party or Patriots and they selected cases simply because the applications had those names in the title. That was wrong, that was absolutely incorrect, insensitive, and inappropriate."

It was later discovered that the question from the ABA audience was actually planted, virtually word for word, by Lerner.[6] The confession was an extraordinarily clumsy attempt at damage control. She wanted to get ahead of the news cycle before the inspector general released a scathing report on the IRS's extraordinary practice of singling out and targeting tea party groups applying for 501(c)(4) tax status in the two years leading up to the 2012 elections.

Activist bureaucrats in an agency of the federal government singling out citizens, based on their political ideology, and effectively impinging upon their political speech. Sounds familiar, doesn't it?

"The other thing that happened was they also, in some cases, sat around for a while," Lerner continued to her ABA audience of tax professionals. "They also sent some letters out that were far too broad, asking questions of these organizations that weren't really necessary for the type of application. In some cases you probably read that they asked for contributor names. That's not appropriate, not usual. . . ."

It was always "they" who were in the wrong. Not "we," or "I."

America would soon discover that nonprofit organization applications that contained the phrases "tea party," "government spending," "government debt," "taxes," "patriots," and "9/12" were isolated from other applications and subjected to extra paperwork and inquiries, delaying some approvals by as much as 1,138 days.[7] Your citizen group's application would have been flagged if you had stated in the IRS application your desire to "make America a better place to live." Targeted groups were instructed to disclose hundreds of pages of private information, including the names of volunteers, donors, and even relatives of volunteers; résumés for each governing group member; printouts of websites and social media contents; and book reports of the clubs' suggested reading materials. Even the content of members' prayers were scrutinized.[8]

According to National Public Radio, of the conservative and libertarian groups requesting tax exempt status in 2012–2013, only 46 percent were approved, with many more never receiving a response from the IRS. In contrast, 100 percent of progressive groups were approved. Additionally, the IRS asked conservative groups an average of 14.9 questions about their applications, but progressive groups were asked only 4.7 questions.[9]

Karen Kenney of the San Fernando Valley (CA) Patriots testified before the House Ways and Means Committee about her experience being targeted by the IRS, that her application for 501(c)(4) status was ignored for two years. Suddenly the IRS demanded an enormous amount of information, including personal information about employees and donors and transcriptions of meetings and candidate forums, allowing them only twenty days to comply.[10]

Dianne Belsom of the Laurens County (SC) Tea Party testified that she was told that she would receive information on her application for 501(c)(4) status within ninety days. More than a year later, she had still heard nothing. Once an election year rolled around, they started bombarding her with requests for information similar to the kinds listed above. After filing all requested information, the IRS asked for more, including repetition of previous requests. At the time of her testimony, her application had been pending for more than three years with no sign of resolution.[11]

Toby Marie Walker of the Waco (TX) Tea Party said that the total number of documents requested from their group by the IRS would have filled "a U-Haul truck of about 20 feet."[12]

POLITICAL SUPPRESSION?

Why so many questions, so many forms? One clue might come from an unrelated article regarding the tax treatment of certain nonprofit university activities. The IRS was crack-

ing down. How? According to a Bloomberg article from
November 2011:

> *Lois Lerner, the IRS's director of tax-exempt organi-*
> *zations who is overseeing the investigation, says many*
> *schools are rethinking how and what they report to the*
> *government. Receiving a thick questionnaire from the*
> *IRS, she says, is a "behavior changer."*[13]

What behavior was the IRS trying to change with regards to citizen groups wanting to make America a better
place to live? Maybe the thick questionnaires and intrusive
inquiries served a particular purpose? Maybe the IRS intended to change behavior? Stan Veuger of the American
Enterprise Institute argues that the IRS effectively suppressed "get-out-the-vote" activity by tea partiers in 2012:

> *The Tea Party movement's huge success [in 2010] was*
> *not the result of a few days of work by an elected official*
> *or two, but involved activists all over the country who*
> *spent the year and a half leading up to the midterm*
> *elections volunteering, organizing, donating, and ral-*
> *lying. Much of these grassroots activities were centered*
> *around 501(c)4s, which according to our research were*
> *an important component of the Tea Party movement*
> *and its rise. The bottom line is that the Tea Party*
> *movement, when properly activated, can generate a*
> *huge number of votes—more votes in 2010, in fact,*
> *than the vote advantage Obama held over Romney in*
> *2012. The data show that, had the Tea Party groups*

continued to grow at the pace seen in 2009 and 2010,
and had their effect on the 2012 vote been similar to
that seen in 2010, they would have brought the Repub-
lican Party as many as 5–8.5 million votes, compared
to Obama's victory margin of 5 million.[14]

The targeting of tea partiers and groups that sought to "make America a better place to live" mattered. Their political activity was suppressed and their First Amendment right to speak and assemble effectively taken from them. Bureaucrats buried them under mountains of questions. Attorney General Eric Holder has promised, in a different context that happened to accrue to President Obama's political advantage in the 2012 campaign, to "not allow political pretexts to disenfranchise American citizens of their most precious right."

Incredibly, Lerner originally maintained that these out-of-control line workers in the agency's Cincinnati office, one of the agency's largest and most significant branch offices, "didn't do this because of any political bias. They did it because they were working together. This was a streamlined way for them to refer to the cases. They didn't have the appropriate level of sensitivity about how this might appear to others and it was just wrong." It was just an innocent mistake made by low-level civil servants—"line staff" based in Cincinnati, Ohio.

Except that it wasn't innocent. And it wasn't limited to low-level staff at the Cincinnati office. The targeting and intimidation of tea party groups started right before the 2010 elections and continued, despite knowledge of the practice by supervisors, all the way up the chain of command, right to the desk of IRS chief counsel William Wilkins, an Obama political appointee.

When summoned to address this issue before the House Oversight Committee, Lerner said, "I have done nothing wrong," before promptly clamming up and refusing to answer any questions on the subject. The Obama administration clammed up as well. President Obama's qualified outrage acknowledged even less than Lerner did in her first admission. Obama apparatchik David Axelrod argued that the "vast" size of the federal government makes it impossible for the president to know what is going on beneath him in the executive branch. Democrats quickly went into attack mode, trashing the inspector general and accusing the chairman of the House Oversight Committee, Darrell Issa, of a "partisan witch hunt."

In Washington's parlance, this is called "spinning."

So what happened to the promise of a better world under the benevolent hand of big government? If you really do believe in the "awesome authority" of the state, wouldn't you be the first in line demanding accountability from those who abused power? The whole spectacle felt more like the actions of a Third World junta, not the executive branch of the United States government.

A HISTORY OF ABUSE

Needless to say, this is not the first time agents at the IRS have picked winners and losers for the benefit of a sitting president, or for the benefit of a zealous bureaucrat. In 1963, having determined that Martin Luther King was "the most dangerous Negro" in America, J. Edgar Hoover set out to destroy him. One of the more powerful tools at the FBI's disposal was the IRS, and the agency's access to confidential data, particularly the donor list of MLK's organization, the Southern Christian Leadership Conference. The FBI "hoped to use the IRS's list of SCLC donors to send them phony SCLC letters warning that the organization was being investigated for tax fraud. This, they hoped, would dry up the funding of King's group and thereby neutralize it." [15] King and the SCLS were both audited by the IRS at Hoover's behest.[16]

During his term as president, John F. Kennedy used the IRS to target conservative nonprofits and other political foes, as well as to obtain the confidential tax information of rich conservatives H. L. Hunt and J. Paul Getty.[17]

Robert Kennedy commissioned a report from labor leader Victor Reuther on "possible administration policies and programs to combat the radical right." The report argued for using the IRS as a weapon. "Action to dam up these funds may be the quickest way to turn the tide." Reuther suggested denial of tax-exempt status and investigations of corporations suspected of being right-wingers. Reuther said: "[T]here is the big question whether [they] are themselves complying with the tax laws," indicating that he may have supported audits against these organizations.[18]

Richard Nixon's presidency ended abruptly for crimes including his willingness to use the IRS to selectively punish his political enemies. Former IRS chief Johnnie Mac Walters reports that under Nixon, he was handed an enemies list of two hundred people and instructed that the White House wanted them "investigated and some put in jail."[19] Nixon, of course, resigned his office when faced with the possibility of impeachment for his crimes, and for repeatedly engaging "in conduct violating the constitutional rights of citizens." According to Article 2 of the Articles of Impeachment:

> *He has, acting personally and through his subordinates and agents, endeavoured to obtain from the Internal Revenue Service, in violation of the constitutional rights of citizens, confidential information contained in income tax returns for purposed* [sic] *not authorized by law, and to cause, in violation of the constitutional rights of citizens, income tax audits or other income tax investigations to be initiated or conducted in a discriminatory manner.*[20]

Historically, IRS abuse seems to follow a pattern. In 2001, the academic journal *Economics & Politics* published an empirical study of IRS audits and concluded that, "Other things being the same, the percentage of tax returns audited by the IRS is markedly lower in states that are important to the sitting president's re-election aspirations. We also find that the IRS is responsive to its oversight committees."[21]

An Offer You Can't Refuse

So, is abusing the power of the IRS just politics as usual? John F. Kennedy and Bill Clinton did it, but so did Richard Nixon and George W. Bush.

Was the IRS just taking orders from President Obama, and from powerful Senate Democrats like Dick Durbin and Max Baucus, all of whom publicly, loudly telegraphed their desire for the IRS to go after certain sinister 501(c)(4)s?

Or is there something even more ominous going on?

There is real evidence that Lois Lerner is a partisan with an ax to grind, and is willing to use her positions of power to advance her personal agenda. In 1996 she used her position as a Federal Elections Commission lawyer to go after Illinois U.S. Senate candidate Al Salvi, a Republican challenging Senator Dick Durbin. Late in the election, Salvi was hit by an FEC complaint filed by the Democratic National Committee, a charge that would dominate the headlines for the remainder of the campaign, which Salvi lost to Durbin. The charges were later dropped in court as frivolous, but not before Lois Lerner put Salvi through a bureaucratic and legal wood chipper.

It started off with an offer Salvi couldn't refuse: "Promise me you'll never run for office again, and we'll drop the case," she told him.

> Salvi said he asked Lerner if she would be willing to put the offer into writing.
>
> "We don't do things that way," Salvi said Lerner replied.

Salvi then asked how such an agreement could be enforced.

According to Salvi, Lerner replied: "You'll find out."

The aspiring Republican never ran against Durbin again. "It was a nightmare," Salvi says now. "Why would anyone run for office again after all that?"[22]

In September 2013, new emails surfaced that directly contradicted the timeline set out in Lerner's original ABA mea culpa. These emails directly rebutted the claim that the targeting of tea partiers was not politically motivated. "Tea Party matter very dangerous," she emailed her staff in February 2010. "Cincy should probably NOT have these cases."[23] Reacting to an NPR article emailed to her by a fellow staffer, titled "Democrats Say Anonymous Donors Unfairly Influencing Senate Races," Lerner responded, "Perhaps the FEC will save the day."[24]

Data suggest that Lerner isn't the only IRS employee with an agenda. According to Tim Carney at the *Washington Examiner,* "IRS employees also gave $67,000 to the PAC of the National Treasury Employees Union, which in turn gave more than 96 percent of its contributions to Democrats. Add the PAC cash to the individual donations and IRS employees favor Democrats 2-to-1." In the Cincinnati office, every political donation made in 2012 by employees went to either Barack Obama's reelection campaign or to liberal Democratic senator Sherrod Brown.[25]

CONCENTRATED BENEFITS AND DISPERSED COSTS

Public choice economists argue that government decisions on how money is spent and who benefits from regulation are driven in large part by the various interests that stand to win or lose. The payoff for successfully influencing the political decision-making process can be highly motivating. These are the "concentrated benefits" that special interests seek when they show up in Washington to lobby. Those who don't show up—the rest of us—don't typically even know that our ox is about to be gored on their table. Even if we did, the cost of showing up and attempting to influence the outcome of legislative horse-trading would be prohibitive. So knowingly or not, we all incur the "dispersed costs" of bigger government.

It's typically less of a goring, and more of a slow bleed. More like a frog in a pot of water slowly brought to boil. You don't really know it's happening. Pennies more for the sugar you buy at the grocery store, or the gradual devaluation of the dollars in your pocket through the Fed's expansion of money and credit supplies. These are just a few of the countless other ways that the incestuous self-dealing of Washington insiders transfers wealth from you to them.

But sometimes the costs are vivid, and people rise up in protest. It's happening more and more as the costs of good, real-time information plummet. It happened in 2008, when America opposed a $700 billion bailout of Wall Street. It is happening again in public opposition to ObamaCare, particularly over the unjust transfer of wealth from younger, poorer Americans to older, wealthier ones. The president's health-care reboot grows increasingly unpopular years after its enactment. People are discovering the hard way that the political promises made to buy the votes needed to pass the massive scheme were mostly expedient lies. More and more people want out of the new government exchanges.

Acting IRS chief Danny Werfel is one of those people. Testifying at a hearing, he told the House Committee on Ways and Means that "I would prefer to stay with the current policy that I'm pleased with rather than go through a change if I don't need to go through that change."[26]

His view is echoed by the National Treasury Employees Union—yes, IRS employees have union representation—who are aggressively lobbying to keep their members out of ObamaCare. Here is the opening paragraph of the letter the unions asked members to send to Capitol Hill:

> *I am a federal employee and one of your constituents. I am very concerned about legislation that has been introduced by Congressman Dave Camp to push federal employees out of the Federal Employees Health Benefits Program (FEHBP) and into the insurance exchanges established under the Affordable Care Act (ACA).*[27]

So they are looking to exempt themselves from the same onerous law that the IRS is enforcing on us? Are you kidding me? What happened to equal treatment under the laws of the land? As outrageous as that sounds, consider this: The IRS commissioner who oversaw the exempt organizations division of the IRS from 2010 to 2012, the very time frame when the agency was targeting conservative and libertarian groups, is now in charge of the new division at the IRS enforcing ObamaCare. Her name is Sarah Hall Ingram, and she directly reported to Lois Lerner.[28]

Ingram and her new army of IRS enforcement agents will be imposing fines on young people who choose not to be conscripted into ObamaCare. But thanks to an Office of Personnel Management (OPM) ruling, IRS employees, other federal employees, and the politicians and their staffs who drafted and enacted ObamaCare are all effectively exempt. Through President Obama's personal request, OPM is allowing members of Congress to retain benefits conferred by the Federal Employees Health Benefits Program, despite the fact that ObamaCare would otherwise require them to purchase the same health insurance programs available to the population at large.[29]

ObamaCare for thee, but not me? Depends on whom you know in Washington.

THE INSIDERS VERSUS AMERICA

What if we have reached a tipping point in America, where the progressive dream of a protected class of civil servants has turned into something else completely? What if the hope of change is really just a big, powerful, selectively abusive, and very expensive nightmare? It used to be well understood, or at least widely believed, that they worked for us. We were taught in high school civics that members of Congress and the president and all government workers were, in fact, employees of We the People.

What if public servants now represent a privileged class,

the most powerful special interest group in the world? Consider this: In 2011, the federal government had 4,403,000 employees.[30] To lend perspective to this number, consider that Wal-Mart has fewer than half this number of employees, coming in at 2.1 million, and McDonald's has only 1.9 million.[31]

Of course, no one believes that the nice lady who greets you when you enter your local Supercenter is out to get you. She is there to help you. That nice lady, and the Wal-Mart she works for, really work for you and your return patronage.

When is the last time you felt that way about a federal government employee?

What's most remarkable about the IRS targeting of conservative and libertarian grassroots organizations is the length and scope of the practice. The discrimination against the IRS's self-categorized "tea party cases" was an agency-wide practice that was discovered, broadly known about, discussed up the chain of command, and continued for years. Watergate was the product of a few bad actors, and the malfeasants were caught, stopped, and brought to justice.

This is a big deal, a potential tipping point where the self-interests of bureaucrats looking to protect their jobs dovetailed nicely with a chief executive, in an election year, looking to protect his job. Think of the implications of the federal government as the largest special interest group in the world.

Do they work for us, or we them?

Becky Gerritson, president of the Wetumpka (AL) Tea Party and a target of the IRS, answered this question un-

equivocally in her unbending testimony before the House
Ways and Means Committee:

> *I am not here today as a serf or a vassal. I am not beg-
> ging my lords for mercy. I am a born-free, American
> woman—wife, mother and citizen—and I'm telling
> MY government that you have forgotten your place. It
> is not your responsibility to look out for my well-being
> or monitor my speech. It is not your right to assert an
> agenda. The posts you occupy exist to preserve Ameri-
> can liberty. You have sworn to perform that duty. And
> you have faltered.*[32]

Becky's testimony "went viral" on YouTube, fueled by
the simple, commonsense values that she personified, values
that I believe still define America.

The American ideal is about your liberty, not their
power.

It's no longer Republican versus Democrat. It's not about
good government or bad government. It's not even "liberal"
versus "conservative." It's about limiting the government's
monopoly on force and unleashing our freedom to try, to
choose, to take responsibility, and to make things better. It
is about the political elites and the insiders they collude with
versus America.

It's Them versus Us, for sure.

CHAPTER 4

GRAY-SUITED SOVIETS

*If you give me six lines written by the
hand of the most honest of men, I will find
something in them which will hang him.*
—CARDINAL RICHELIEU[1]

IT'S NOT PARANOIA IF they really are out to get you.

The Internal Revenue Service systematically targets its
critics: average American citizens simply trying to comply
with complex laws, and simply exercising their First Amendment voice in the public debate.

The attorney general authorizes wiretaps on the phones
of reporters at the Associated Press. The National Security
Agency spies on you, presuming your guilt until proven
otherwise.

The Orwellian-named Affordable Care Act (neither affordable nor caring) will collect all of your personal data,
from an alphabet soup of federal agencies including the IRS,
DHS, and DoD—even your private health insurance infor-

mation. All of this "private" information will be centralized in a sweeping government "data hub" housed in the Department of Health and Human Services.

The elected members of Congress and their staffs who drafted and enacted the ACA—better known as ObamaCare—and the career "civil servants" at the IRS, NSA, and HHS—all of them to be trusted with so much discretionary power over your life and information about you—are seeking to exempt themselves from the same laws they will impose on us.

President Barack Obama, a Democrat seemingly impatient that anyone would question *his* administration's handling of *our* privacy, tells us "it's important to recognize that you can't have 100 percent security and also then have 100 percent privacy and zero inconvenience."[2]

Republican senator Lindsey Graham, with casual disregard for your Fourth Amendment guarantee to privacy, assures us on all the government's snooping: If you have nothing to hide, then you "don't have anything to worry about."[3]

SUBDUING ALL SPHERES

In *Human Action,* Ludwig von Mises reminds us that government is the "opposite of liberty." Government always means "coercion and compulsion." We shouldn't be surprised by the arrogant dismissals of President Obama, Senator Graham, IRS officials, ObamaCare bureaucrats, or anyone else who would assert their power over our freedoms. As Mises notes, "It is in the nature of the men handling the apparatus of compulsion and coercion to overrate its power to work, and to strive at subduing all spheres of human life to its immediate influence."[4]

I would suggest two corollaries to Mises's observed "compulsion and coercion": complexity and control. The reason you want a simple set of rules that are applied equally across the board is precisely that the monopoly power of the state is dangerous. Combined with complex and intrusive laws, a government monopoly on power puts incredible authority into the hands of faceless, gray-suited bureaucrats with ideological axes to grind, self-interests to protect, and personal scores to settle. Think J. Edgar Hoover, or Lois Lerner. Think gray-suited soviets imbued with an agenda quite contrary to yours thumbing through your financial and health history records, fishing for some discrepancy to get you with.

What could possibly go wrong?

If you want to see the corrosive effects of unfettered discretionary power imposed from the top down—unfireable employees in the executive branch with an ability to target or ignore, choose winners or losers based on ideology or personal agendas—you need look no further than Washington, D.C., today.

WITH THE PRESIDENT'S DECISION to illegally rewrite ObamaCare in real time, arbitrarily delaying implementation of the employer mandate until after the 2014 elections,[5] it is clear who will really decide future health-care decisions.

And it's not you. The influential interests in the big business community successfully lobbied for a(nother) delay. No such luck for the rest of us. We are expected to comply.

Of course, the government's blank checkbook comes with a complex web of laws that have grown more convoluted over time, the rules rearranged by various special interests, and then again by the bureaucrats inside the enforcement agencies. This political give-and-take has little to do with your interests. The growth of governmental infrastructure represents a political equilibrium that is anything but economically efficient for the rest of us.

I call it the Complexity Industrial Complex. The more complicated things get, the better off the insiders are. Bureaucrats feed on complexity, a permanent rationale for expanded budgets and higher compensation, and a fat meal ticket back outside the government as a highly compensated guide to corporations looking to navigate the labyrinth of laws and rules. Navigate, or, more likely, exploit to their advantage.

Take the authors of the Patient Protection and Affordable Care Act, for example. Since drafting that legislation, more than thirty of its architects have found lucrative positions lobbying for deep-pocketed corporations such as Delta Air Lines, Coca-Cola, and British Petroleum, charging big bucks to help companies navigate the regulatory mazes they helped create.[6]

Incumbent corporations—"big business"—often lobby for, and get, new complexity as a strategy to keep underfunded upstart competitors out of the market. In reality, monopoly market power is typically the by-product of this unholy collusion between complexity-mongers in and outside government. Market share can only be protected permanently in partnership with the power monopolists inside government.

Complexity is also the skirt that power abusers hide behind. One of the excuses used by partisan Democrats in defense of Lois Lerner and other IRS employees who targeted activist groups based on ideology was the convolution of campaign finance laws. It's an interesting argument coming from the congressional architects of campaign finance regulations. They, of course, wrote the First Amendment–gagging laws that give so much latitude to the permanent bureaucracy. Not one of these congressional authors of convoluted campaign finance laws has ever called for scrapping the whole monstrous structure in favor of a simple defense of First Amendment rights to political speech.

In an op-ed in the *Washington Post,* Elijah Cummings, the ranking Democrat on the House Oversight Committee, wrote: "The sad reality is that while House Republicans have devoted time and taxpayer money to attempting to smear the White House, they have failed to examine part of the underlying problem the IRS faces: inadequate guidance on how to process applications of organizations seeking tax-exempt status."[7]

When the IRS can't understand its own rules or properly follow the law because of "inadequate guidance," that ought to be a warning sign that the system is broken in Washington.

CODE RED

There is, of course, a method to their madness. Consider the remarkably corrupt federal tax code.

In 2013, the federal tax code was a whopping 73,954 pages long,[8] or about four million words.[9] To put this in perspective, consider that the Guinness world record holder for the longest novel, Marcel Proust's seven-volume *In Search of Lost Time*, is less than a third of this length.[10] Anyone attempting to read the tax code had better have plenty of time on their hands. At a rate of fifty pages a day, it would take more than four years to make it through the whole thing.

Obviously, no one can reasonably be expected to know all the rules and regulations buried in this unfathomable tome. And that may be precisely the point. Even people who are paid to enforce these rules don't understand them. Former IRS commissioner Douglas Shulman freely admitted in an interview on C-SPAN that he does not file his own taxes, explaining that, "I find [hiring a tax preparer] convenient, and I find the tax code complex."[11]

When the head of an agency is incapable of understanding the rules he is charged with enforcing, something seems fundamentally wrong. It is estimated that Americans spend 6.1 billion hours a year simply complying with the tax code,[12] and the accounting costs of compliance total between $67 billion and $378 billion every year. Imagine all the good that could be accomplished if that time were instead spent on productive activity. The Mercatus Center has estimated that the total loss to the economy resulting from our complex tax code is as much as $609 billion a year, and this does not include the time and money spent by lobbyists to petition for special tax treatment.[13]

This is a far cry from the way the founding fathers initially envisioned the system of taxation for their new country. For more than a hundred years after its founding, the United States government was funded purely with tariffs, excises, and receipts from the sale of federally owned lands. Direct taxes, such as the now-familiar income tax, were out of the question.

Sadly, this simpler state of affairs was not to last. The Civil War brought an unprecedented level of expenses, and new taxes were an easy way to collect revenue in a hurry. The Revenue Act of 1861 established the Internal Revenue Service and created the first incarnation of what we would recognize as the modern income tax.

This was originally intended to be a "temporary" measure to finance the war, but history has shown that there are few things more difficult than ending a temporary government program. In 1913, Congress ratified the Sixteenth Amendment to the Constitution, making the individual income tax a permanent feature of law. Then, the top tax rate was less than 1 percent, and the tax code totaled twenty-seven pages in length.[14]

Within minutes of enactment, moneyed insiders began rewriting the income tax code to carve out exceptions that favored their interests.

In a way, tax code complexity is inevitable. Special interests seeking exclusions, deductions, subsidies, and refunds have a strong incentive to lobby the government to advance their cause. If they know someone, if they have juice in the

Capitol, a new provision is added and the tax code becomes more of a labyrinth requiring experts, inside and out. I've personally heard business lobbyists make their case to congressional staffers. Every dollar in "tax expenditures" will generate a threefold return for the government, they will claim. They always have the "blue-chip" study under their arm to back it up.

When I worked on Capitol Hill in the 1990s, we tried to repeal the federal sugar program. Big Sugar lobbyists descended upon Washington like a swarm of locusts, distributing backslaps, PAC checks, and free coffee cups that claimed that the sugar growers' subsidy cost the federal government "Zero." Still got the cup. And Big Sugar still has its special deal. In fiscal year 2013, the program that was not supposed to cost you anything cost taxpayers $280 million dollars. The tab will increase in 2014, according to the *Wall Street Journal*.[15]

It's not so hard to cook up a case for special treatment that sounds persuasive to eager ears. But there's always a *quo* in exchange for *quid*.

A simpler tax code benefits all of us, but the incentives for individuals to act to protect the principle of equal treatment under the law are small by comparison. Concentrated benefits for the insiders, and dispersed costs for the rest of us.

Far more important than all of the wasted time resulting from tax code compliance, these complex and difficult-to-understand rules fuel abusive and discriminatory practices. No one can manage absolute compliance with this Byzantine mountain of tax regulations, meaning that anyone is vulnerable to punishment at the discretion of the IRS.

Complexity plus compulsion means you—"the most honest of men"—are vulnerable to someone else's agenda.

In the 1960s, the IRS targeted Martin Luther King Jr. because he was deemed a threat to various government interests. Most recently, the IRS targeted moms who want to "make America a better place to live." What assurances do you have that someday you won't get sideways with some government agent? Will someone with power you don't have target you for speaking up against "Common Core," a top-down set of education standards that take still more say away from parents? Will you even know who that gray-suited bureaucrat is who decides that you are "dangerous"?

This problem does not end with the tax code. All areas of law have become so complex that full compliance is impossible. The potential for selective persecution is obvious. The Environmental Protection Agency, for instance, has itself acknowledged that complete compliance with the Clean Air Act creates an administrative burden that is "absurd" and "impossible." We are literally drowning in an ocean of dictates and orders, so much so that it has been estimated that the average American breaks three federal laws every day without realizing it.[16]

The fact that all of us are continually breaking laws we don't even know exist, combined with legal standards that do not consider ignorance an excuse for noncompliance, means that any of us could be punished at any moment, entirely at the discretion of the enforcers. Obscure, rarely enforced laws can easily turn into tools of oppression when selectively applied for political reasons.

The law is meant to be an instrument of protection for

the people, not a tool of arbitrary and discriminatory punishment. When even the enforcers cannot keep up with the sheer scale of our legal code, there is nothing to stop our protectors from turning into oppressors.

FIRST DO NO HARM

Which is one very good reason to oppose a government takeover of our health care.

What was once a simple relationship between patient and doctor has become a tangled morass of regulations and middlemen, equally damaging to Americans' health and wallets alike.

ObamaCare promised to fix these problems by creating a government-controlled system and a dramatic expansion of government-funded health insurance. Despite the persistent unpopularity of the proposal, President Obama devoted the bulk of his first term in office to pushing the bill through Congress. Defending the law from grassroots critics has dominated his second term's domestic agenda, as the systematic problems with it create real chaos.

Imagine what will happen when the new middleman between you and your doctor is a career civil servant who likely has an agenda contrary to your own. Perhaps this faceless decider determines whether or not your son or your wife or your mom gets the procedure they need to live. It might be someone else's "cost-benefit" analysis that deems your needs unaffordable. Maybe a gray-suited bureaucrat at HHS decides to poke through your political activities in the

Data Hub first. Maybe there just isn't enough money in the centralized system to cover all of the demands from all of the sons and daughters and grandmas and grandpas who were dumped into it by their employers. How to choose? Who are the winners and losers? Will you get a say, an appeal if the answer is no?

Of course not. ObamaCare is less about the quality and affordability of your health care, and more about who controls your health-care future. ObamaCare seeks to fix a system caused by too much government, by injecting still more government discretionary control into the system.

In the early part of the twentieth century, there was no third-party payment system for health care. Individuals who got sick would pay their doctors directly, and this personal accountability kept costs low; doctors could not make money if none of their patients could afford to see them.

All of this began to change in the run-up to World War II. In a politically motivated, economically illiterate effort to boost employment and reduce income inequality, President Franklin Delano Roosevelt imposed wage controls on businesses, dictating the amounts they were allowed to pay their workers.[17] Good economics always takes a backseat in the ordering of political priorities. As John Maynard Keynes once said: "In the long run, we're all dead." He was speaking to the political class, who would eventually hear instead: "In the long run, when the political reckoning for the consequences of shortsighted policies come, I'll be out of office, maybe lobbying Congress for more complexity that favors my clients."

So it was with FDR and wage and price controls, and his

penchant for throwing bad economic ideas against the wall to see how long they would stick.

Business owners naturally wanted to employ the best talent they could find, but these wage controls hampered their ability to attract the best workers by offering higher salaries than their competitors. In order to get around the law, and to appease one of his most important political constituencies, the labor unions, FDR authorized nonwage perks, one of which was the first incarnation of employer-sponsored health insurance.

Why link health insurance to your employment? It doesn't make much sense, and it gives someone else more control over your health-care decisions. This government-driven market distortion was the first of many corruptions that put distance between the patient and the doctor. With deep-pocketed businesses picking up the bill for their employees, doctors had much less incentive to keep costs down, forcing insurance premiums to rise and making health care largely unaffordable for anyone outside the system. The tax code further complicates the problem with its unequal and confusing treatment of health-care costs. Inside the politically structured system, health insurance benefits were tax-free.

But outside the system, you pay inflated costs with after-tax dollars. This is what amounts to political "compassion." It sounds so good when read compellingly from the dais teleprompter, but in practice the little guy gets screwed.

Instead of addressing these issues directly, ObamaCare doubles down on an already corrupted system. Imposing a fiendishly complex new system, against the will of the people, is no way to resolve the fundamental problems with the market for health services. If doctors and insurance companies raised prices when businesses were paying, they will only raise them more when government, with no profit motive or competition to restrain its spending, is picking up the tab. Congress and the administration will, in turn, impose price controls on hospitals and doctors. Because it worked so well when FDR tried it.

Rising costs and rising demand for "free" services can only lead to one outcome. It's called rationing of health-care services. Or "death panels." Or some gray-suited soviet, who just learned that you once sent $250 to Ron Paul's 2012 presidential campaign, choosing winners and losers in a very complicated system that no one really understands.

One of the key features of the law is to require insurance companies to accept patients with pre-existing conditions, a requirement that undermines the entire concept of insurance. Insurance exists as a safeguard against a possible future disaster, something that is unforeseeable but potentially devastating. You cannot insure against something that has already happened. You cannot buy homeowner's insurance on a house that is already on fire. If you could, it would no longer be insurance but a mere shifting of costs from one person to another.

Since insurance companies can now be certain that every one of their clients will be filing a claim, their rates will have to rise to compensate for the increased costs. Hence, the "individual mandate" targeting, by force, young, healthy people who cannot afford and don't need the government-defined plans. Much more on the "social injustice" of this later.

The fact that ObamaCare will fail in its aims of reducing costs and increasing coverage seems to have become common knowledge, with huge numbers of businesses lobbying to get out of the employer mandate. And the federal government has been happy to comply.

The Department of Health and Human Services has announced that about 1,200 businesses have been granted exemptions from the ObamaCare employer mandate.[18] Labor unions are not happy with the law, either, and have sought waivers en masse for their membership. Thus far, labor unions representing 543,812 workers and private companies employing 69,813 workers have been granted waivers.[19] If ObamaCare is supposed to be such a good deal for workers, why do so many of them want out of it?

Private companies are scrambling to deal with the increased costs of the law as well. UPS has announced that it will no longer provide health coverage for employee spouses, while Walgreens and IBM are dumping employees from their employer-sponsored plans, asking them to buy private health insurance instead.[20] As of this writing, five million Americans have received cancellation notices from their insurance companies,[21] and some estimates put future cancellations as high as a hundred million.[22] So much for Obama's original promise that "if you like your health insurance, you can keep it."

Of course, the political class, fresh off a closed-door lobbying campaign to protect their generous health insurance plans did get to "keep it." President Obama responded to their demands, personally asking that the Office of Personnel Management allow members of Congress to retain the massive subsidies conferred by the Federal Employees Health Benefits Program, despite the fact that ObamaCare would otherwise require them to purchase the same health insurance programs available to the population at large.[23] The OPM, controlled by the president, quickly solved the problem for the insiders. According to the sympathetic *Washington Post*:

> [L]awmakers and their staffs previously had about 70 percent of their insurance premiums underwritten by the federal government through the Federal Employees Health Benefits Program. . . . Under pressure from Congress, the Office of Personnel [created a new ruling] saying the federal government could still contribute to health-care premiums.
>
> The final rule would keep the subsidy in place only for members of Congress and affected staff who enroll in a Small Business Health Options Program (SHOP) plan available in the District of Columbia. Such plans most commonly will be aimed at employees of businesses with fewer than 50 workers, but perhaps the theory is that each lawmaker and his or her staff constitute a small business.[24]

When Senator David Vitter (R-LA) introduced an amendment to eliminate this *de facto* exemption for congressional employees, Democrats descended on him in a rage, calling the effort "mean-spirited," while Republican staffers quietly lobbied against the efforts behind the scenes to preserve their special treatment.[25] Insiders on both sides of the aisle are equally invested in this two-tiered system, the inequities of which will surely come back to haunt them as "consumers" in the ObamaCare exchanges look for relief from the new system's sticker shock.

Due to disorganization and a general reluctance for anyone to comply with an obviously bad law, many of ObamaCare's deadlines have been delayed. Nearly two-thirds of U.S. states have outright refused to set up the health-care exchanges required by the law, forcing the federal government's hand and resulting in multiple pushbacks of the initial deadline. The employer mandate has also been delayed until 2015, due to panicking businesses realizing that they were unprepared to bear the full financial toll of the requirement. Cuts to Medicare and numerous eligibility requirements for health insurance subsidies have also been put on hold.[26] Overall, the rollout of the president's signature legislation has been nothing short of a chaotic mess.

So what we have is a large, cumbersome, unworkable, ineffective health-care program that nobody wants, but which is nevertheless the law of the land. What agency could be trusted to enforce such a disastrous policy? You guessed it: the IRS.

Of all the federal agencies that could possibly be tapped to implement ObamaCare, it would be hard to come up with a worse choice than the IRS. We have already established the political corruption to which the agency is susceptible, but there are a number of other reasons that entrusting them with our health care is a uniquely bad idea.

First, ObamaCare adds a total of forty-seven new duties and enforcement powers to the agency, which has admitted to lacking the necessary resources to fulfill even its existing duties. In a hearing defending the IRS's discriminatory practices, then IRS commissioner Stephen Miller testified that "it would be good to have a little budget that would allow us to get more than the number of people we have."[27] How can we expect fair and equal treatment from an agency that blames its ethical violations on lack of funding?

The gray-suited soviets at the IRS have proven time and again that they have no respect for the privacy of individuals or their records. Apart from the prying questions asked of tea party groups in the discrimination scandal, the IRS has allegedly violated the law by seizing 60 million medical records from a California health-care provider.[28] Allowing the agency to manage health-care subsidies and impose penalties will open the door to further abuse and remove the control of sensitive personal information from individual patient-doctor relationships.

Finally, the IRS has no expertise in the field of health care. They are being asked to regulate an industry they know nothing about, and given the quality of IRS employees we have seen in the public spotlight lately, it seems overly optimistic to expect them to be a quick study, or unbiased enforcer. Indeed, the sample of officials paraded before congressional committees in recent months exposes a workforce that is seemingly immune to public oversight or controls.

In a misconduct hearing, Miller responded to an alarming number of questions by claiming that he didn't know, couldn't remember, or wasn't sure of the answers. He expressed no knowledge of his own employees and claimed to lack any opinion on whether the actions of his agency were appropriate.[29]

If the IRS is truly as disorganized and unknowing as its leaders claim, why on earth should the agency be allowed to handle your health care? The other option, of course, is that they are obfuscating, parrying with political opponents to run out the clock, knowing that career civil servants will be around longer than any single politician or even a president.

This alternative scenario hardly seems reassuring. It raises a fundamental question about the legitimacy of the way business is conducted in the federal labyrinth. Do they work for us? Are they accountable to We the People? Or us to them?

In 2009, then–House Speaker Nancy Pelosi infamously said, "*We have to pass the bill so that you can find out what is in it*, away from the fog of the controversy" (emphasis added).

Well, now we have passed it, and the law is far more unpopular than the proposed legislation ever was. When the Healthcare.gov website debuted to disastrous malfunction on October 1, 2013, it became evident that this unpopularity was not simply a result of the "fog of controversy," but an intuitive understanding of government ineptness.

Even after the botched launch, however, Nancy Pelosi was still cheerleading the law, oblivious to the reality that was quickly gripping the rest of the country. Speaking several weeks after the website launch, Pelosi insisted that she wanted to "say every chance I get how proud we are of [ObamaCare.]" She then went on to make the tenuous claim that the law "is life, a healthier life, liberty to pursue your happiness, as our founders promised."[30]

It is telling that the president who consistently thinks of himself as the smartest guy in the room appeared downright baffled by his administration's inability to successfully remake the nation's entire health-care system with an unbending belief in smarter government. Redesigning one-sixth of the American economy, he now concedes, is more complicated than he imagined:

> But even if we get the hardware and software working exactly the way it's supposed to with relatively minor glitches, what we're also discovering is that insurance is complicated to buy. And another mistake that we made, I think, was underestimating the difficulties of people purchasing insurance online and shopping for a lot of options with a lot of costs and lot of different benefits and plans and somehow expecting that that would be very smooth, and then they've also got to try to apply for tax credits on the website.[31]

An embarrassingly botched website is just the beginning. Friedrich Hayek refers to the grandiose pretensions of government redesigners as a "fatal conceit," because of the unforeseen, and often dire, consequences of big government designs on private life. Real people are getting hurt by the pretensions of ObamaCare, and the only real winners seem to be the insiders who will administer the new complex structure.

ObamaCare seeks to supplant a broken system with more of the meddling and discretionary reengineering that broke it in the first place. Increased complexity and difficulty of compliance is precisely the opposite of what is needed to fix health care in America. The only certainty is that more bureaucrats will be hired, and that they will be given extraordinary discretionary power over the health care of your family.

AN EMPIRE OF DATABASES

If turning over your medical records to the IRS sounds scary, it is nothing compared to the immense Federal Data Services Hub the Obama administration has planned. Wary of the decentralization of information, the president has announced his plans to collect a massive amount of personal data on every citizen, stored in one place and overseen entirely by the infinite wisdom of career bureaucrats who are virtually unfirable.

As part of the Patient Protection and Affordable Care Act, the data hub is designed to allow health-care exchanges to access personal information on patients through the IRS, the Social Security Administration, the Department of Homeland Security, the Veterans Health Administration, the Department of Defense, the Office of Personnel Management, and the Peace Corps.[32] Why do health-care exchanges need so much information? It has to do with the complex eligibility requirements for the various health-care subsidies included under the law. Since these subsidies are determined by how much money you make, exchanges need access to your tax records, as well as any other information that could qualify you for certain benefits, or indeed penalties.

One of the major problems with having all of this sensitive information in one place is that any successful attempt to break into the Hub by an outside party could result in the identity thefts of millions of Americans. Think about it: names, email addresses, telephone numbers, Social Security numbers, tax data, health insurance records, immigration status, and prison records will all be available for one ingenious hacker to take and use how he will.[33]

But surely the government would not allow such a thing to happen. The safeguards on such a repository must be enormous, right? Actually, the Obama administration has already missed numerous deadlines in implementing security measures for the Hub. Although the law requires that these safeguards be in place, the administration has so far been unable to meet its own standards. The Identity Theft Resource Center reports 34.1 percent of all data breaches in 2013 were related to health care.[34] This is not exactly a reassuring thought. Can we really trust all the details of our private lives to an organization that has consistently failed to fulfill its promises?

The government's record on keeping its information secure is not exactly exemplary. An inspector general's report found that the IRS accidentally disclosed the confidential taxpayer information of thousands of people from 2009 to 2010.[35] The Social Security Administration has mistakenly disclosed thousands of names, birth dates, and Social Security numbers.[36] And in 2012, a lone hacker managed to obtain 3.6 million names and Social Security numbers from a South Carolina database.[37]

The problem is compounded by the fact that the incentives for information theft will be greater than ever before. The payoffs for a malicious identity thief would be exponentially greater than when data was stored separately across a wide variety of individually encrypted databases. Democratic representative Jackie Speier of California expressed concern over this, saying that the Hub would have a "bull's-eye" on it for hackers.[38]

But the threat that this data could be obtained by someone outside the government may well be overshadowed by the potential for internal abuse of the data by government employees. The recent IRS and NSA scandals make it plain that a simple security clearance does not alleviate the temptation to abuse one's authority. On the contrary, as the editors of the *Wall Street Journal* have pointed out, "putting the IRS in charge of a political program inevitably makes the IRS more political."[39] The more personal information we allow these agencies to have, the easier it will be for them to identify, and potentially target, their political enemies. In the light of recent events, this is a danger that we should all take very seriously. The simple fact is that the ObamaCare data hub will eliminate any semblance of privacy we have as far as the federal government is concerned, and any overzealous employee will be able to wreak havoc with the lives of ordinary Americans.

The regulatory notice detailing the particulars of the

Data Hub says that the government is free to disclose any of the information it has collected to a variety of individuals and agencies without the consent of the individual. This information sharing is not limited to secure government agencies, but includes "contractors, consultants, or grantees," as well as law enforcement officials.[40] So not only will your data be collected and shared within government departments; it can potentially be dispersed to any number of private contractors without your knowledge. The notice insists that the data will only go to those people who need it for their records, but it is not clear at what level the "need to know" threshold will be set. It is easy to envision a situation in which the unscrupulous are afforded easy access to sensitive documents.

Like the botched development of Healthcare.gov, the Obama administration rushed to hire a slew of "patient navigators," individuals whose jobs consist of helping others sign up for the ObamaCare exchanges, a step that would involve the collection of a great deal of personal information. Rather than requiring the same kind of security clearances or background checks necessary for positions in sensitive government agencies such as the FBI or the IRS, the Department of Health and Human Services waived any such requirement, instead asking only for a twenty- to thirty-hour online training seminar. A high school diploma is not even required.[41] Among the groups eager to take up positions as "patient navigators" are Planned Parenthood, senior citizen advocacy organizations, and churches.[42]

By now, both the potential for abuse and the seriousness of the consequences should be obvious. If the Obama administration is prepared to allow barely trained, agenda-driven workers from off the street access to your most private data, is there reason to believe that there will be any serious effort to protect the rights of enrollees on the exchanges? Enrolling in the ObamaCare exchanges means surrendering all of your most private information into the hands of a government that has proven irresponsible, untrustworthy, insecure, and indiscreet at every turn.

But go ahead and trust them; they're from the government.

A Perfect Storm

The inability of citizens to comply with the labyrinthine laws of their country, the imposition of an oppressive and ineffective health-care scheme on an unwilling public, and the revelation that we have no privacy from our government and little recourse if accused of a crime have coalesced to expose the excesses of a government out of control. This dark cloud is a call to action for those who wish to preserve their freedoms and liberate themselves from an increasingly oppressive federal bureaucracy.

The trend toward more power in Washington, D.C., runs headlong into a world that is quickly trending in the opposite direction. The Internet and its ubiquitous social media mutations are quickly disrupting and mercilessly dismantling many of the outdated, top-down institutional structures that used to tell us what our choices were from a predetermined set of options. Now we are free to choose, to shop, to gather information, to organize, to vote, and to associate as we please based on our own preferences.

This collision is imminent. Like an incoming cold front rolling forward on a hot summer day, this is a perfect storm between the power hoarders in the Halls of Discretion and your right to design, as best you see fit, your own future.

One way or another, something's going to give.

CHAPTER 5

SAME AS THE OLD BOSS

IN JANUARY 1973, RICHARD Nixon ended the military draft in the wake of a series of high-profile draft-card-burning protests by antiwar activists. (That's right, a Republican ended the military draft. And it was *Nixon*.) His presidency would soon enough end ignominiously, though, in part due to his eagerness to use the IRS to selectively punish his political enemies. The Democrats, the Republicans, the left, and the press were all outraged by this remarkable abuse of executive power.

The current IRS scandal, where the agency systematically targeted moms organizing their communities to defend constitutional principles like the freedom to associate and peaceably assemble, elicits no such outrage from Democrats or the many tentacles of leftist activist organizations. Few seem willing, or even interested in, defending *everyone's* civil rights and the First Amendment protection of political speech for *those guys*. How sad.

The Democrats' and progressives' act of omission on IRS harassment leading up to the 2012 election is bad enough. Don't they remember Watergate? Are they no longer repulsed by what federal agents at the CIA and the IRS did in an all-out bureaucratic onslaught to silence Dr. King? But then there is the left's blatantly partisan act of commission as head cheerleaders for a new individual mandate that involuntarily conscripts young people into ObamaCare, whether they like it or not. They are literally drafting millennials into a system designed by administration technocrats, powerful committee chairmen, and a whoring mob of big insurance interests that got to the table first to carve out an acceptable return on their political investments.

Meet the new boss. Same as the old boss, but worse.

Advocates of conscripting our youth into ObamaCare typically hide behind the fact that various advocates on the "right"—notably Mitt Romney and Newt Gingrich—advocated on behalf of the individual mandate. It's a ridiculous argument for them to make, because you know that they would oppose, lockstep, this sort of reverse Robin Hood scheme if it were proposed by a President Romney or a President Gingrich.

Why not apply a consistent set of principles, consistently applied, regardless of which party label is attached?

It's a dirty business, and this oppressive wealth transfer from young Americans to special interests and the more-wealthy appears to be the Achilles' heel of the new, insanely authoritarian progressive movement. Whatever the clarion call of "social justice" was supposed to entail, surely garnishing the wages of the young and struggling to bolster the earnings reports of Big Insurance and Fortune 500 dinosaurs was never part of the plan.

DEAD IN THE LONG RUN

While the generational theft inherent in ObamaCare will become increasingly obvious as young people sit down and consider their coerced "choices," the relentless process of making financial commitments we can't afford today, to be foisted upon the buckling shoulders of future taxpayers tomorrow, is pretty much business as usual in Washington, D.C.

In his important critique of modern public finance practices, the late Nobel Prize–winning economist James Buchanan referred to the dominance of rob-the-cradle fiscal policies as the sad legacy of John Maynard Keynes. Our democracy was in deficit, he said, literally and structurally. Keynes, who single-handedly severed the cord between Adam Smith and the new "macroeconomics," was culpable.

Since America's founding, it was generally understood that governments should not spend money they don't have. "What is prudence in the conduct of every private family," Smith argued in *The Wealth of Nations,* "can scarce be folly in that of a great kingdom." Under the old rules, says Buchanan, "government should not place future generations in bondage by deficit financing of public outlays designed to provide temporary and short-lived benefits." But all that changed with the publication of Keynes's *General Theory of Employment, Interest, and Money*. Here's how Buchanan puts it:

> *With the completion of the Keynesian revolution, these time-tested principles of fiscal responsibility were consigned to the heap of superstitious nostrums that once stifled enlightened political-fiscal activism. Keynesianism stood the Smithian analogy on its head. The stress was placed on the differences rather than the similarities between a family and the state, and notably with respect to principles of prudent fiscal conduct. The state was no longer to be conceived in the image of the family, and the rules of prudent fiscal conduct differed dramatically as between the two institutions. The message of Keynesianism might be summarized as: What is folly in the conduct of a private family may be prudence in the conduct of the affairs of a great nation.*[1]

So Keynes provided a pretense of intellectual legitimacy to the natural, and very destructive, instincts of politicians wanting to spend more money than public coffers held. "In the long run we are all dead,"[2] was how Keynes himself rationalized the idea of spending binges that were unsustainable. Such shortsighted thinking fits perfectly into the psyche of politicians thinking in two-year increments, or presidents thinking about a single reelection bid after four years in office.

Keynes is, in fact, dead, but if you are eighteen years old today, you are on the hook for his bad economics. The irresponsible choices being made today will have to be paid for in the future, by our children and grandchildren. You can only pass the buck so far. As of this writing, every man, woman, and child in America owes more than fifty thousand dollars toward the national debt, a number that grows larger with every passing day.[3] With a declining birth rate and a generation of baby boomers starting to enter retirement, Social Security has become neither social nor secure. It is a massive, cross-generational wealth transfer, with millennials losing a huge chunk of their paycheck to pay into a system that they know is unlikely to exist when they seek to retire in forty years. The finances for Medicare are far worse, and you can only expect the financial burden on young people to grow as ObamaCare raids Medicare coffers.

It all seems crazy, and young people are rightfully cynical about their futures.

On top of all of this, President Obama's controversial health-care scheme imposes its "individual mandate," Washington-speak for an incredibly regressive "tax" imposed on young, healthy people that forces them to buy mandated health insurance plans that they can't afford and don't need, or else pay a fine. With all of the new federal add-ons, insurance company lobbyists insisted on a coercive means of making young people cross-subsidize the benefits of older, wealthier patients. Imagine being able to force new customers to overpay for your product. What a deal.

And why not? If you were the CEO of McDonald's and the White House offered you a chance to mandate the consumption of Big Macs for lunch, would you take the deal, strings attached? For a number of health-care and big business interests lobbying their parochial agenda in the battle over ObamaCare in 2009, the answer, apparently, has been a hearty "Yes, We Can."

And so it was that ObamaCare would include a mandate, with the full coercive force of the federal government and the IRS behind it, that eighteen-year-olds buy health insurance from a politically prescribed set of plans approved by the federal government. Insurance providers just love it when healthy young people sign up for gold-plated insurance plans. It's because young people seldom use the health-care system.

Which is why so many young people choose to go uninsured in today's screwed-up health-care marketplace. It's a question of costs versus benefits on a really tight budget.

Why would progressives flack for such an affront to the cause of social justice? Why would they ever support a political shakedown that fattens the bottom lines of the federal government and insurance kingpins alike? It just doesn't make any sense.

No sense at all, until you understand that the individual mandate is the heart of the ObamaCare redistribution scheme. Without overcharging young people, the ObamaCare exchanges simply won't work.

And this is what the new left, Democrats, and progressives are foisting onto already overburdened youth. When did the American left decide that it was cool to subsidize the Man on the backs of millennials struggling with unprecedented student loan debt and a jobless recession that never ends? Just think about the real burdens foisted upon American youth by the failed experiments in big government.

It Sucks to Be Young

The most important event in anyone's personal launch into orbit is finding a good job. If you are graduating from high school or college in 2014, this is no simple task. A market crash in 2008 enabled by easy money, easy credit, easy spending, and an expanding federal bureaucratic expanse was met with more of the same, effectively creating economic refugees of the millennial generation. The hits keep coming, and the affront of an individual mandate adds to the weight of twenty-somethings already pig-piled by the secondary effects of a big, intrusive, government agenda paid for with borrowed money.

The so-called recovery from the Great Recession has been especially hard on the young. Rates of new job creation have not returned to precrash levels, or even acceptable average rates. Deceptive decreases in the unemployment rate have been driven by ever lower levels of participation in the labor force. By September 2013 this rate had plummeted to its lowest level in thirty-five years.[4]

In other words, more people are giving up.

Unemployment has remained a serious problem for everyone, but those between the ages of twenty and twenty-four have been hit the hardest. The lack of workplace experience among recent graduates means that they have the least leverage with which to negotiate with employers, and when jobs have to be cut, theirs are the first to go. As of August 2013 the unemployment rate among Americans aged twenty to twenty-four was a dismal 13 percent,[5] and a Pew Research Study found that a record number—36 percent—

of millennials are forced to live under their parents' roofs.[6]

Adding insult to injury is the insane cost of higher education. More and more young people graduate with a degree that is effectively worth less than the debt they accrued to finance it. The total amount of student loan debt has ballooned in recent years, now reaching an astonishing $1 trillion, an average of more than $24,000 per student.[7] A third of this debt has been incurred to support advanced degrees, more paper pursued by more students as a means to delay entering the uninviting job market.[8] Student loan debt tripled between 2004 and 2012 and was the only kind of household debt that continued rising throughout the Great Recession. In 2012, nearly one-third of student loan borrowers were delinquent in their loan repayments.[9]

Why are students accruing so much debt? Because the costs of higher education, fed by bureaucracy and government subsidies, have skyrocketed to the point where few students are able to afford them. Here again, notice the bitter irony of federal intervention and borrowed dollars: The result is even less affordable education. From 2012 to 2013, the net cost of one year at a public, in-state university averaged more than $12,000.[10] And that's not counting the opportunity cost of missing out on four years of real-world work experience.

This situation is clearly unsustainable. Like all financial bubbles, this one must inevitably burst, bringing with it still more economic chaos. The Federal Advisory Council has warned that the rapid growth in student loan debt looks eerily similar to the housing bubble that precipitated the

Great Recession.[11] As the Obama administration continues to push the idea that everyone should go to college, with financial help from new federal subsidies, the resulting increase in costs and debt issuance could well lead to another crash as students are forced to default.

Students cannot afford the escalating costs of college and graduate school, and the debt they are racking up today will dog them throughout their adult lives. The cushy, cloistered lives of college professors look not unlike those of career federal bureaucrats. The permanent administrative class in our public university system is bleeding young people dry, all for an overpriced education that will not pay for itself, even if you can find a job.

It sucks to be young today. You have inherited all of the financial burdens accrued by older generations and bad fiscal management in Washington, D.C., but you see few of the opportunities once virtually guaranteed by the American dream.

Millennial voters, of course, swung big for Obama in both 2008 and 2012.[12] In 2008, 18- to 29-year-olds broke for Obama 66 percent, to 32 percent for his challenger, John McCain. In 2012, 60 percent of voting young people supported the president's reelection, compared to 37 percent for Mitt Romney. Their aversion to the crusty candidates the Republican Party has offered up makes some sense, I suppose. The research surrounding recent presidential and midterm primaries and caucuses suggests that support for the Republican brand is nearing a historic low. What is the reason for this?

Here's the popular hypothesis: Nationally, America is

experiencing growing racial and ethnic diversity, which is concentrated among younger voters. This benefits Democrats. Race, ethnicity, and age are strongly associated with one's opinions about government. It's a safe bet that these factors account for greater degrees of liberalism within the younger age groups.

Much-hyped autopsies confirm the obvious. The "Grand Old Party" stresses neither "Grandness" nor the idea of "Party." They underscore the "Old." Consensus is, they're outmoded, out-of-touch, aging white guys. This won't come as breaking news, but Democrats are perceived as more tolerant.

Now things get more complicated.

A 2013 poll conducted by the polling company for FreedomWorks found that attitudes among young people are shifting toward a preference for smaller government.[13]

> We asked young voters to weigh whether "you would favor a smaller government with few services but lower taxes, or a larger government which provides more services but has higher taxes?" Asked in this reflection-of-reality way, Millennials' views on the role of government flip. A majority of young voters favor "smaller government with fewer services but lower taxes." College-age and recent graduates (ages 18–24) favor smaller government by 51 to 45 percent. Young voters ages 25–32, who have been on the job market for a few years and are more likely to be paying taxes, favor smaller government by 64 to 24 percent.

These numbers signal a reversal of the trend observed in a 2010 study by the Pew Research Center, which found that 53 percent of millennials agree with the statement "government should do more." [14]

Why is this the case? Well, the Pew report suggests ethnic and racial diversity does play a part. However, these youngsters are also less religious, less likely to have served in the military, and will probably emerge as the most educated generation in American history. That's partly because we've matured to meet the "the demands of a knowledge-based economy," but it also has something to do with the fact that young people grasp at degrees because they can't find jobs.

While they remain largely optimistic, their emergence into adulthood has been stunted. In the wake of the Great Recession, careers are hard to come by, and young people's first jobs are often low-paid and unappealing. While they remain more upbeat about their own economic futures, and the overall health of our nation, that optimism has been tempered.

INSANELY AUTHORITARIAN

And then if you do find a job, even in this economy, the IRS comes knocking on your door to make sure you "voluntarily" comply with the ObamaCare revolution. They will make you an offer you can't possibly refuse.

The payoff for youth's heightened political allegiance to the Hope and Change Agenda is a coerced payout to cross-subsidize the government health insurance plans of those who are older and wealthier. The whole ObamaCare scheme would collapse unless carried forward on the backs of the young, and the successful implementation of an "individual mandate" that forces them to buy expensive health plans.

What would Jerry Rubin, the Yippie war protestor, do? He would torch his ObamaCare card without a moment's hesitation, chanting, "Hell no, we won't go."

This is an opportunity. Or, as the president would say, "a teachable moment." It seems like a unique time for peaceful civil disobedience and noncompliance—a grassroots rejection of Washington's corrupt ways, where insiders win at the expense of the rest of America. I thought it would be cool to skewer progressives for their hypocrisy on the individual mandate by tapping into a bit of 1960s hippie zeitgeist. I am after all, a seasoned Deadhead with nearly one hundred live Grateful Dead shows under my belt. So my colleagues at FreedomWorks got together with two of the best liberty-minded student groups—Students For Liberty and Young Americans For Liberty—and organized a series of "Burn Your ObamaCare Card" protests channeling the ethos of antiwar protestors. At least David A. Graham at *The Atlantic* gets the joke:

> *This is an ingenious cultural appropriation. On the one hand, FreedomWorks is drawing a pointed link between protests against the Vietnam-era draft—a hated government program that depended on forcing the country's young to sign up for something not in*

their best interests—and the Affordable Care Act, a hated government program that depends on forcing the country's young to sign up for something that's (arguably) not in their best interests. There are, to be sure, some pretty serious differences. . . . But there's a smirk behind it all too. FreedomWorks is taking a treasured image of the anti-war left, the high-water mark of American progressive political action, and seeking to make it the right's own.[15]

I would have thought that the individual mandate would have united civil and economic libertarians—left, right, center, and anyone with a sense of fairness—against the insider trading of the political class in Washington, D.C. Instead, the progressive standard-bearers at *Mother Jones* went all apoplectic at the very notion of dodging the ObamaCare draft. "So not only [is FreedomWorks] going to be encouraging people to break the law," froths *Mother Jones,* "they're literally going to be encouraging people not to buy health insurance. . . . It's times like this that words fail those of us with a few remaining vestiges of human decency."[16]

REALLY? *THE VERY FEW last remaining vestiges of human dignity?* What exactly is undignified about young people making rational economic choices regardless of what their Washington, D.C., overlords might deem best for them? Maybe young people are fed up with being bled dry, unable to save, to build their own dream, and wanting to be free from someone else's grand plan.

I read panic in *Mother Jones*'s unhinged rant—a real-ization that the American left is now the Man, forcing its authoritarian plans on an unwilling generation at the point of a gun. How ironic.

Secretary of Health and Human Services Kathleen Sebelius called FreedomWorks' ceremonial card-burning efforts "dismal." [17] Her comment came before the actual rollout of Healthcare.gov, an embarrassing failure which truly defined the term. HHS is so worried about its ability to conscript enough young, able bodies into the master plan that it rolled out a $700 million corporate PR campaign to convince twenty-somethings to buy an overpriced Edsel. [18]

The problem with the Obama administration's costly propaganda, to paraphrase Bill Clinton's memorable line from the 2012 Democratic National Convention, is sim-ple "arithmetic." Hypotheticals and political talking points have been replaced with actual price tags based on actual options. And the numbers are indeed "dismal."

According to one front-page *Wall Street Journal* article, "The success of the new health-care law rides in large mea-sure on whether young, healthy people . . . decide to give up a chunk of disposable income to pay for insurance." [19]

Jonathan Scarboro, when asked, did the math. "I'm not going to pay for that," he says of mandated coverage. At thirty years old, Scarboro makes $29,000 a year and is now required to pay, at minimum, $147 a month, with a $6,350 deductible. "It breaks down to: Can I afford it? And, am I getting my money's worth?"

All good questions, Jonathan, and the answer is "no." Better to opt out of the program, burn your "ObamaCare card," and pay the fine. That's the conclusion of eight out of ten young people interviewed by the *Journal*. Turns out, this strategy could save young Americans hundreds of dollars a year. A National Center for Public Policy Research study recently revealed that single, childless Americans between the ages of eighteen and thirty-four could save at least $500 by opting out of ObamaCare and paying the $95 individual mandate penalty instead.[20]

The Obama administration made its case for the Affordable Care Act by insisting that the law would reduce health-care costs and premiums, but this claim turns out to be just as true as the president's memorable promise that you could keep your current plan, period. Particularly for young people, the very purpose of the Affordable Care Act is to increase costs and coverage. When the final prices for the plans being offered on health-care exchanges were announced, the news was far from reassuring for a generation struggling to keep their heads above water as it is. While prices vary considerably by state, the national average cost to a young, healthy person turned out to be $163 a month.[21] That's for the lowest-quality "bronze" plan they are permitted to buy. The kind of coverage young people actually might need, insurance against catastrophic injuries, was not an option. After all, a plan like that would not have the redistributive effect so desired by the architects of ObamaCare. The rational choice for young, healthy people is *not to comply*. A study from the American Action Forum concluded that, on average, premiums for men under thirty will increase by 260 percent.[22]

What eighteen-year-old can afford to pay $163 a

month—$2,000 a year—for health insurance that covers services most will never use? How can we, in good conscience, impose this cost on millennials who can barely make ends meet now?

Reason magazine's Nick Gillespie sums up the wholesale generational theft of ObamaCare nicely:

> *It's a feature and not a bug of the President's signature health care law that insurance premiums for those under 30 are likely to increase significantly to allow premiums for older Americans to fall. Indeed, the whole plan hinges on getting 2.7 million whippersnappers out of a total of 7 million enrollees to sign up in the individual market during the first year. If too many older and sicker folks flood the market, the system will crash even faster than the HealthCare.gov website.*[23]

There is a rational alternative to this government-run health-care hostage situation. A better, patient-centered model would cut out all of the gray-suited middle men who currently corrupt the effective provision of health care.

Why not respect young people enough as sovereign individuals to let them choose? Why not let young people save for their future health care needs tax free in exchange for voluntarily choosing a catastrophic health insurance policy? Notice that, without really trying, I just solved two of the main challenges in any health reform plan: portability and pre-existing conditions. Turns out that choice, individual savings, and personal responsibility work well, even in health care. The only problem with my plan is the political class's loss of control over you. It turns out that independent young people can't be bought as easily on Election Day.

In the meantime, young people will do the simple arithmetic and reject ObamaCare, perhaps going without. Or they may choose to buy outside the government-engineered system.

The Man says he has a plan for you. Better for young people to turn on, tune in, and drop out, and take back for themselves control of their own health-care needs, back from the insanely authoritarian new left.

A Lost Generation

It has become fashionable to stereotype the new generation of Americans as narcissistic, disconnected, and lazy. Last year, *Time* featured a cover branding them "The Me Me Me Generation."[24] This is partially a manifestation of every generation's belief that "kids today" just aren't as good as they used to be, but the mainstream attitude toward millennials has acquired a vitriol as puzzling as it is unjustified. No wonder politicians struggle to connect with today's youth.

In his book *Invisible: How Millennials Are Changing the Way We Sell,* author T. Scott Gross confronts some of these myths. He points out that millennials are reluctant to buy into tradition for the sake of tradition, that they prefer participation to observation, and that they embrace diversity in a way earlier generations never have.[25] These are not the values of the social parasite who prefers government dependence to individual initiative.

Millennials are not disconnected, they are just lost, looking for something better. They're searching for a political home. They are a generation without a voice, sold out by the Democrats they helped put in office, and uninspired by the limp and disingenuous Republican alternative. The poor performance in presidential elections by establishment candidates like John McCain and Mitt Romney shows that the same old ideas are not going to win over younger voters. They are tired of endless wars, tired of broken promises, and tired of politics as usual.

Barack Obama got elected by claiming to be a new kind of politician. He was young. He was energetic. He was super cool. He looked different than the pasty, old Washington insiders we had gotten so used to. He spoke with charisma and enthusiasm for his cause, and he utilized new technologies to reach out to young people in a language they understood.

But it all turned out to be total BS. Obama promised the end of lobbyists, but he employs an army of them. He promised to run the most transparent administration in history, but the Committee to Protect Journalists reports that his administration's efforts to control the media are "the most aggressive . . . since the Nixon administration." "This is the most closed, control freak administration I've ever covered," said David E. Sanger, veteran chief Washington correspondent of the *New York Times*.[26]

Obama's presidency has been rocked by scandal after scandal. From Fast and Furious to IRS abuses and unprec-

edented cyber-snooping at the NSA, secretive insider tactics
have been business as usual for the last five years.

He campaigned on a platform of peace, but he has conducted military operations in multiple countries at once,
costing American lives and racking up still more debt.
He has even ordered drone strikes on American citizens
without granting them the due process guaranteed by the
Constitution.

The nation's youth are tired of having their hopes dashed
by broken promises. They are searching for something new,
something different. Public opinion polls are beginning to
reflect this desire for a change. A poll by Harvard University
of 18- to 29-year-olds finds that trust for every aspect of
government, from the Supreme Court to the presidency, is
declining, and a growing number disagree with the idea that
government spending can cause greater economic growth.[27]
A 2013 Rasmussen Reports survey found that 63 percent
of respondents think a government with too much power is
more dangerous than one with too little power, the highest
number ever recorded.[28]

The president recognizes that he is in trouble with
young people. The core principle of his health-care law is
that the young will have to buy plans they don't want or
need to subsidize older Americans. How do you convince
an underemployed eighteen-year-old that it is their social
responsibility to pick up the tab for their grandparents? To
address this problem, the president did what he always does:
He gave a speech.

Speaking at a so-called "youth summit" at the White

House in December 2013, Obama attempted to browbeat a crowd of 160 young activists into compliance, urging them to return to the troubled Healthcare.gov website and sign up for ObamaCare.

> *Look, I do remember what it is like being twenty-seven or twenty-eight, and aside from the occasional basketball injury, most of the time I kind of felt like I had nothing to worry about. Of course that's what most people think until they have something to worry about. But at that point, often times, it's too late. And sometimes in this debate, what we've heard are people saying, well, I don't need this, I don't want this; why are you impinging on my freedom to do whatever I want.*[29]

Unable to win them over with talk of social responsibility and their shared sacrifice, the president instead resorted to using fear to convince people to support a program that more and more were finding unpalatable. *Look: You just might die without "free" preventative care.*

According to a December 2013 study released by the Institute of Politics at Harvard University (IOP), kids today just aren't buying what Obama is selling anymore. A majority under 25 would throw Obama out of office given the chance. Fifty-seven percent of millennials now oppose ObamaCare. Among the most coveted potential enrollees currently without health insurance, fewer than one third of 18- to 29-year-olds plan to enlist in the ObamaCare exchanges.

That's a sea change from the salad days of hope and change.

The survey, part of a unique thirteen-year study of the attitudes of young adults, finds that America's rising generation is worried about its future, disillusioned with the U.S. political system, strongly opposed to the government's domestic surveillance apparatus, and drifting away from both major parties. "Young Americans hold the president, Congress and the federal government in less esteem almost by the day, and the level of engagement they are having in politics are also on the decline," reads the IOP's analysis of its poll. "Millennials are losing touch with government and its programs because they believe government is losing touch with them." [30]

In 2011, a CNN poll that has been conducted regularly since 1993 found a record high number of respondents thinking like libertarians, with 63 percent saying that government is doing too much and 50 percent saying the government should not favor a particular set of values. [31] Another CNN poll asked, "Do you think the federal government has become so large and powerful that it poses an immediate threat to the rights and freedoms of ordinary citizens, or not?" Sixty-two percent of respondents answered yes, it does pose a threat, up from 56 percent in 2010, the last time that question was asked. [32]

A polling company survey asking about the role of government found the highest levels of support for libertarian values in more than a decade. [33] Another found growing levels of support for libertarian ideas among the Republican Party. [34]

Young people are often characterized as economically conservative and socially liberal. A better configuration, or at least a challenge of old, broken premises, might be a clear-eyed skepticism regarding the wisdom of giving third parties the power to make decisions for us.

On questions like the definition of marriage, a better solution might be to let individuals and communities and proven religious institutions decide for themselves. Social norms are created by people working together, not by governments. Governments, and the political process, and the inevitable self-interested agendas that define political outcomes, typically corrupt our best social traditions. America's youth have never been defined by conformity and submissiveness. You don't have to agree with the choices of others. You just shouldn't use force to make them conform to your own set of values.

As long as you don't hurt people, or take their stuff.

NOT THE PREFERRED NOMENCLATURE?

Conservatives, as you might understand the usage of the term, used to be "liberal," as in pro-freedom of the individual and pro-limiting the power of the state. Now, many of us use the term "classical liberal." Former socialists in Europe, prefer to call us "neo-liberals." Today's liberals in the U.S. used to be "progressives" in the mold of Teddy Roosevelt and the splinter Bull Moose Party of 1912, but have chosen to misappropriate our classic "liberal" brand. The modern left has so trashed the meaning of "liberal" that they have re-appropriated "progressive" as their preferred nomenclature. "Neo-conservatives" used to be socialists, and despite their respect for traditional social values they still cling to the socialist penchant to rearrange things and manipulate the choices we would otherwise make for ourselves.

Nobody wants to be branded a socialist, or a fascist, or a communist in the United States anymore, including the president. "I am not a socialist," Obama pointedly told an editorial board at the *New York Times* in 2009. It appears to bug him enough that he reiterated the distinction at the *Wall Street Journal*'s 2013 CEO Counsel meeting: "People call me a socialist sometimes, but you've got to meet real socialists, you'll have a real sense of what a socialist is."[35]

Is it all clear enough for you?

Skeptics of too much government power—right, left, and center—struggle with brands. And maybe that's natural. Maybe this is the inevitable lot of individualists. We don't always want to be categorized, or collated into one of the two preexisting mail slots that say "Republican" or "Democrat."

In 1856, the Republican Party replaced a Whig Party that had lost its philosophical bearings to the point of being an empty shell. It had once stood against tyranny and a too powerful executive branch. Today's Republican Party in many ways is suffering from a political identity crisis of its own, and has failed too many times to deliver on its message of limited government and individual liberty. Democrats are more reliably authoritarian, now controlled by a progressive ideology, always wanting more government involvement in our lives.

Some Republicans, typically incumbents-for-life who have gotten way too cozy with the power and special relationships with the lobbying class that come with it, have lost credibility, often selling out their principles to special interests and the preservation of their own political skins. The Democrats have the very same problem, but have done even worse as the party in control, expanding military intervention in foreign lands, abandoning their promised commitment to civil liberties in favor of the cult of personality that is Barack Obama. And then there is the reverse Robin Hood scheme called ObamaCare.

The old way of doing business isn't going to cut it anymore. Regardless of the brand name, it's pretty clear that millennials are up for grabs, looking for something better than just a new, hipper boss in Washington.

CHAPTER 6

THE RIGHT TO KNOW

THE INTERNET CHANGES EVERYTHING.

In a free society, voluntary cooperation based on mutually beneficial choices and agreements helps individual people to get along and prosper, to not hurt other people or take their stuff. This is how it is possible for millions of people with very different goals and personal beliefs and private knowledge to come together to create things so much greater and more complex than any one person could have done alone.

Don Lavoie, my favorite professor at George Mason University, argued that this freedom-based model creates "a greater social intelligence" that cannot be replicated or reverse-engineered by the most sophisticated planning by the smartest among us. Lavoie got the basis of many of his ideas from Friedrich Hayek. Hayek's work on economic coordination was a critique of various attempts by governments to plan our activities from the top down. Why did government planning typically fail? Because knowledge about what people want and need is not something that can simply be aggregated minus the process of free people figuring things out. This is the process that we all go through, sorting out the infinite pieces of information that bombard each of us in our daily lives. Through our choices, based on our personal knowledge, a pattern emerges that helps others who don't know anything about us know what they need to know to meet our demands. Hayek, of course, got many of his ideas from Ludwig von Mises, who in turn drew from Carl Menger and Scottish Enlightenment thinkers like Adam Smith and Andrew Ferguson.

Writing in the 1760s, Ferguson anticipates the wisdom of crowds:

> *The crowd of mankind, are directed in their establishments and measures, by the circumstances in which they are placed; and seldom are turned from their way, to follow the plan of any single projector.*
>
> *Every step and every movement of the multitude, even in what are termed enlightened ages, are made with equal blindness to the future; and nations stumble upon establishments, which are indeed the result of human action, but not the execution of any human design.*[1]

Advances in our knowledge about how civil society works come from a type of intellectual cooperation not unlike the process of entrepreneurship—part creative thinking and part listening and learning from others who know more than you do. Sometimes you're the leader, and sometimes you follow the lead. Just like John Coltrane studying his mentor Miles Davis and then breaking the "rules" of jazz, redefining them, making jazz better. Just like Rush ignoring their record label and giving their fans something different, something better.

This push and pull between the creative quest of individuals and the best-understood ways of doing things and institutions that we know work is what Hayekians call the "spontaneous order." I call it beautiful chaos, the constant rearranging of preferences and needs in real time that celebrates the dignity of people and their potential to define, for themselves, a better path in life.

Of course, the Internet changes everything. Everything that worked before based on local knowledge, and freedom, and the ability of people to figure things out, learn from others, and build civil societies, is magnified by the Internet, because it reduces barriers to act, and know, and cooperate.

The Internet also changes the old rules of politics. Smart mobs and crowdsourcing and morphing communities built on social media have all democratized political action and broken down the top-down controls of political parties and the old equilibrium of interest groups that controlled them. Likewise, the old media cartels have been undermined—some might argue mortally wounded—by bloggers and Twitter queens and citizen journalists with smartphone video cameras. We citizens can connect, find out what Washington is up to in real time, and act, all in ways that are becoming easier and cheaper. Concerned moms with tens of thousands of Facebook friends can beat deep-pocketed interests in ways that would have been inconceivable just a few years ago.

Freedom is all about sorting information and distributing knowledge. Politics, the distribution of power, is all about controlling the free flow of information under a pretense of knowledge. The Internet changes this dismal calculus, and cuts out middlemen with hidden agendas. No longer are a few people with tremendous political power able to control the distribution of information about the decisions that are made about the things that really matter, things that impact your life and your stuff, like the taxes you pay, or the health care you are allowed to buy, or even the things you are allowed to say in the public square.

This is a very good thing.

KEEPING AN EYE ON YOU

Unfortunately, the Internet's same liberating forces—the ones that are freeing people—are being leveraged by the government to violate your personal privacy and your liberties. John Perry Barlow, the lyricist for my beloved Grateful Dead and a founder of the Electronic Frontier Foundation, puts this dilemma succinctly: "I have known, ever since I encountered the Internet, that it was both the most liberating tool I had ever seen for humanity, and the best system for extremely granular surveillance that had ever been devised, and that it would always be that way. And that there was always going to be, throughout my lifetime, a battle between the forces of openness and connection, and freedom from repression, and the forces of secrecy and repression. . . ." [2]

The Obama administration, beyond anyone's wildest expectations, has led the charge in this Brave New World of government cyber-surveillance. Their aspiring reach seems to know no bounds. It's a game of hide-and-seek, where yesterday's denials are revised and extended to cover up the latest exposed executive branch tyranny with the false promise of future security. "The national security operations, generally, have one purpose and that is to make sure the American people are safe and that I'm making good decisions," Barack Obama explained to the American people on October 28, 2013. "I'm the final user of all the intelligence that they gather," says the commander in chief. "We give them policy direction, but what we've seen over the last several years is their capacities continue to develop and expand, and that's why I'm initiating now a review to make sure that what they're able to do, doesn't necessarily mean what they should be doing." [3]

In part because of the president's tendency to say "I didn't know" in response to any executive branch abuse-of-power scandal, more people are wondering if he is in charge of the executive branch at all. He "generally" knows what the National Security Agency and other intelligence-gathering functions of the federal government are up to, he says. But he didn't seem to know that the NSA was listening to German chancellor Angela Merkel's cell phone calls.

What if the power is now with faceless bureaucrats, not the president? If the president knows the "general" purposes of federal snooping, do you wonder what the extraordinary ones are? Wouldn't you like to know? Given the extraordinary power of the federal government in the era of Big Data, should we trust faceless, unelected bureaucrats with the extraordinary discretionary power to choose you as their next target?

Think about the abuses of power big and small, from J. Edgar Hoover, to Richard Nixon, to Lois Lerner. Think about the qualified, and ever-evolving, promises made by Barack Obama. Does the federal government of the United States have the right to snoop on you, tracking your phone calls and reading your emails? Does Washington, D.C., have the power to limit your speech, spy on the press, or suppress the opinions of bloggers? Does the president of the United States have the discretionary authority to assassinate American citizens on American soil without due process, before guilt is determined in a court of law? Don't you have a right to know?

The president has continuously claimed, responding to a seemingly endless series of revelations that disprove the previous assurances from the White House and various federal agencies, that lines were not crossed, that our constitutional rights were not breached, that your civil liberties were not violated. I don't know about you, but I am not reassured. In fact, I'm certain that things are out of control, and that the balance between our essential liberties and the national security apparatus is fundamentally off, in favor of faceless bureaucrats that we hope are doing the right thing with all that power.

As Americans, our freedoms are broad and our rights

are protected under the Constitution. The government's powers, on the other hand, are supposed to be well defined and strictly limited. But you have to know your rights and vigilantly defend them from the natural tendency of governments to grab power and grow capabilities. Unfortunately, fewer and fewer of us are taught in public school about our guaranteed individual rights. Fewer still take the time to find out and understand the rules for themselves.

This is a very bad trend, and it's our bad if we don't know, or care to know. "Thomas Jefferson often insisted that the ultimate guardians of our rights and liberties are We The People," says the great civil liberties activist Nat Hentoff. "But when many Americans are largely ignorant of the Constitution, an imperial president—like George W. Bush or Barack Obama—can increasingly invade our privacy; and now, with ObamaCare, ration our health care and—for some—our very lives."[4]

BACKASSWARDS

Think he's exaggerating? Consider some of the more extreme views recently expressed by Lindsey Graham of South Carolina. "If I thought censoring the mail was necessary," he told stunned reporters on June 11, 2013, "I would suggest it, but I don't think it is."[5] Graham sits on the Senate Judiciary Committee, with its jurisdiction overseeing "civil liberties." In other words, you could argue, he has a unique responsibility to protect your constitutional rights.

But he doesn't seem to be doing a very good job. In an

interview on *Fox and Friends,* Graham defended the NSA's warrantless surveillance of American civilians, telling the show's hosts, "I don't think you're talking to the terrorists. I know you're not. I know I'm not. So we don't have anything to worry about."[6] He went on to tell the astonished hosts that he was "glad" the warrantless surveillance activity was happening in the NSA.

Guilty until proven innocent? You don't need to be a constitutional lawyer to know that this is backasswards. Graham's view, though not all that unique, is a fundamental inversion of the American concept of justice.

On June 5, 2013, the British newspaper the *Guardian* broke a story about the NSA collecting phone records from millions of Americans who use Verizon.[7] The source of the information was Edward Snowden, a young computer analyst consulting for the NSA. A day later, the *Washington Post*[8] and another *Guardian* story[9] revealed that the surveillance extended to Internet companies as well, enabling the NSA to access emails, photos, videos, and pretty much anything else stored on supposedly secure servers.

Senators Saxby Chambliss (R-GA) and Dianne Feinstein (D-CA) revealed during a hearing of the Senate Intelligence Committee that this type of surveillance had been going on unnoticed for seven years.[10] Feinstein actually defended the program by claiming that the NSA needed access to people's phone records "in case they became terrorist suspects in the future."

The Obama administration jumped right out of the gate with a defense of the NSA, claiming that no personal information or conversational content was being collected,[11] but this was in direct contradiction to a statement made several months earlier at a congressional hearing by Director of National Intelligence James Clapper. Clapper went on record with the following exchange:

SENATOR RON WYDEN: "Does the NSA collect any type of data at all on millions or hundreds of millions of Americans?"

CLAPPER: "No, sir . . . Not wittingly. There are cases where they could inadvertently perhaps, collect, but not, not wittingly." [12]

President Obama continued to deny the accusations of domestic spying with a number of public statements as the story made national headlines. "Nobody is listening to your phone calls," he assured us during a June 10, 2013, press conference.[13] This was followed by an appearance on the *Tonight Show,* where he assured host Jay Leno, "There is no spying on Americans." [14]

But it wasn't true. This assurance, like the often made promise about being able to keep your existing health insurance—period—was a tortured exercise in political expedience. As the weeks rolled on, more information came out revealing the extent of the NSA's spying on American citizens. In addition to more than three thousand supposedly unintentional privacy violations in a one-year period, it was also revealed that a number of NSA employees had admitted to using the surveillance program to spy on former love interests.[15]

So if a bureaucrat with an almost unlimited surveillance tool kit wants to cyber-stalk his former girlfriend, what assurances, besides the president's ever-evolving one, do you have that someone's not stalking you?

All of this was just too much to take, even for the *New York Times* editorial board, hardly charter members of the Ron Paul Revolution: "The administration has lost all credibility on the issue," opined the *Times*. "Mr. Obama is proving the truism that the executive branch will use any power it is given and very likely abuse it."[16]

Critics have rightly pointed out that the passage of the Patriot Act, by a bipartisan majority in a Republican-controlled Congress, unleashed this torrent of domestic snooping. The national tragedy of the September 11, 2001, terrorist attacks had provided the perfect opportunity to extend the reach of government authority in a way that was both public and popular at the time. Promises of safeguards were made, although few who voted for it had actually read the bill. The advocates of broader authorities for the surveillance saw an opportunity to do something undoable before, and they took it. Remember the words of President Obama's former chief of staff, Rahm Emanuel? "You never let a serious crisis go to waste."[17]

KNOW YOUR RIGHTS

The Bill of Rights constitutes the bedrock of our legal protections from the abuse of government power. But fear and apathy and carelessness have started to erode these protections. In the last year alone we have seen egregious violations of the First, Fourth, and Fifth Amendments. It's past time you knew your rights. You and I will have to get involved if we hope to keep them.

The First Amendment is almost universally known. It guarantees freedom of speech and of the press, as well as freedom of religion:

> *Congress shall make no law respecting an establishment of religion, or prohibiting the free exercise thereof; or abridging the freedom of speech, or of the press; or the right of the people peaceably to assemble, and to petition the Government for a redress of grievances.*

The Fourth Amendment is a little less well known, but it is equally important to a free society. It states:

> *The right of the people to be secure in their persons, houses, papers, and effects, against unreasonable searches and seizures, shall not be violated, and no Warrants shall issue, but upon probable cause, supported by Oath or affirmation, and particularly describing the place to be searched, and the persons or things to be seized.*

The surreal scene surrounding the 2013 Boston Marathon bombing revealed how tenuous our hold on these rights might be. The city was thrown into a panic when, on April 15, 2013, two bombs were detonated near the race's finish line, killing three and injuring more than two hundred bystanders. It was a horrific act. After one of the suspects, a nineteen-year-old naturalized citizen, named Dzhokhar Tsarnaev, disappeared into a suburban neighborhood in Watertown, heavily armed SWAT teams embarked on a massive manhunt, barging into private residences and ordering civilians to leave their homes.[18]

The tactic of using fear to chip away at our civil liberties is certainly nothing new. President Obama has adopted the same essential talking points that are always invoked by the defenders of a more powerful government: "You can't have 100 percent security and also have 100 percent privacy."[19] In the real world, of course, we will never see "100 percent security." We live in a dangerous world. Things can happen that are outside our control, and the false promise of perfect safety could easily translate into a blank check for power mongers and a guaranteed path to tyranny.

Representative Jim Sensenbrenner (R-WI), one of the chief architects of the Patriot Act, seems to have come to terms with the unintended consequences of his good intentions.[20] In response to Representative Peter King's (R-NY) assertion that the NSA had acted appropriately 99.99 percent of the time, Sensenbrenner was unequivocal:

> *I don't think 99.99 percent is good enough when you have a court ruling a program unconstitutional in violating the Fourth Amendment and that program had been going on for many months and the NSA violating court orders. It's the court that is supposed to protect the constitutional rights of Americans. I think that James Madison did a pretty good job when he put together the*

*Bill of Rights. I view the Bill of Rights as a sacred doc-
ument and one of the documents that makes America
so much more different than any other country in the
history of the world.*

The Fifth Amendment will be familiar to many due to
the self-incrimination clause and the phrase "pleading the
Fifth," but the actual text contains a number of other important rights as well:

*No person shall be . . . deprived of life, liberty, or prop-
erty, without due process of law; nor shall private property be taken for public use, without just compensation.*

The due process clause has particular resonance today,
after we learned that the Obama administration had ordered the deaths of at least four American citizens through
the use of drone strikes without a trial.

The use of secret courts to circumvent the due process
clause is also alarmingly common, as evidenced by the court
system set up under the Foreign Intelligence Surveillance
Act, commonly known as FISA. The FISA court was created in 1978 as a response to Richard Nixon's increasingly
paranoid efforts to break the law and spy on his political
opponents. The idea was that the federal government would
have to obtain a special warrant from the FISA court before being permitted to conduct domestic espionage operations targeting its own citizens, hopefully putting a stop
to the kind of illegal activities engaged in by the Nixon
administration.

"The constitutional standard for all search warrants is probable cause of *crime*," argues Judge Andrew Napolitano.

> *FISA, however, established a new, different and lesser standard—thus unconstitutional on its face since Congress is bound by, and cannot change, the Constitution—of probable cause of status. The status was that of an agent of a foreign power. So, under FISA, the feds needed to demonstrate to a secret court only that a non-American physically present in the U.S., perhaps under the guise of a student, diplomat or embassy janitor, was really an agent of a foreign power, and the demonstration of that agency alone was sufficient to authorize a search warrant to listen to the agent's telephone calls or read his mail. Over time, the requirement of status as a foreign agent was modified to status as a foreign person.*[21]

The important thing to remember about the FISA court is that the opinions it issues are secret, and that means no public oversight or accountability. The *Guardian* released documents showing that the FISA court had extended more or less blanket authority to the NSA to independently determine which citizens would be targeted for surveillance.[22] The court also gave the NSA broad permissions to store and make use of personal data, even when data was "inadvertently acquired."

A secret court issuing secret permits to a secret agency to spy on American citizens with impunity, effectively operating outside of the law and the Constitution. What could go wrong?

Taking a Stand

It is always true that, once breached, it is very difficult to restore essential liberties and the promised limits on federal power. Each new dollar and every expansion of authority creates a political constituency that wants still more money and authority. That's why governments seem to inexorably grow, not shrink, and new powers created by a Republican Congress are later expanded by a new Democratic president.

That is precisely the speeding train that Senator Rand Paul stepped in front of on the morning of March 6, 2013. Does the president of the United States have the discretionary authority to assassinate American citizens on American soil without due process, before guilt is determined in a court of law? It's a good question, one that deserves a clear answer. It's also a question that Senator Rand Paul had asked of the Obama administration and its chief law enforcement officer, Attorney General Eric Holder, a number of times in 2013. But Paul couldn't get a straight answer. Holder's heavily lawyered nonresponse gave all civil libertarians—left, right, and center—a serious case of the heebie-jeebies.

The first written response the attorney general's office sent to Paul was arrogant, dismissive, and sloppy, seemingly uninterested in fundamental constitutional questions and the constitutionally delineated responsibilities of the legislative branch to check unfettered executive branch power: "The question you have posed is therefore entirely hypothetical, unlikely to occur, and one we hope no President will ever have to confront," wrote Holder. "It is possible, I suppose, to imagine an extraordinary circumstance in which it would be necessary and appropriate for the President to authorize the military to use lethal force within the territory of the United States."[23]

"Hypothetical." "Unlikely." "It is possible." In other words, it's totally up to the president's discretion, and you, Senator Paul, should stop asking questions and mind your place.

This should have been the end of this particular debate, and there was reason to believe that the White House would once again get away with trespassing constitutional boundaries with little debate and even less accountability. This, after all, was a pattern. So when Rand Paul took to the well of the Senate floor to filibuster, effectively stopping Senate legislative business in protest to the administration's nonanswer, there was little reason to believe that he could make a difference. At least that's what the conventional wisdom inside the Beltway believed.

> *I will speak until I can no longer speak. I will speak as long as it takes, until the alarm is sounded from coast to coast that our Constitution is important, that your rights to trial by jury are precious, that no American should be killed by a drone on American soil without first being charged with a crime, without first being found to be guilty by a court.*[24]

The Senate was preparing to vote on the confirmation of President Obama's nominee for director of the CIA, John Brennan. It was likely a surprise when Paul stood up at 11:45 A.M. to address the marble-walled chamber, so few Beltway reporters took any notice. "Certain things rise above partisanship," Paul told the mostly empty room. "And I think your right to be secure in your person, the right to be secure in your liberty, the right to be tried by a jury of your peers—these are things that are so important and rise to such a level that we shouldn't give up on them easily. And I don't see this battle as a partisan battle at all. I don't see this as Republicans versus Democrats. I would be here if there were a Republican president doing this."

The D.C. establishment interpreted Paul's gesture with its typical cynicism. MSNBC host Lawrence O'Donnell won the prize for most blindly partisan and hateful reaction, arguing—if you can call it actual argumentation—that supporting Paul's protest was the very worst decision you could ever make. "If you want to #StandWithRand," O'Donnell asked, "do you want to stand with all of the vile spewing madness that came out of that crazy person's mouth?"[25] After a while, it became clear that he was projecting:

"Horribly flawed."

"Empty headed."

"A little bit more than crazy."

"Performance art."

"Spewing infantile fantasies."

"Sleazier."

"Stark raving mad."

"Psychopath."

OK. Thanks, Larry. I think we got the point. You should get help. Soon.

Senator John McCain came off only slightly more balanced than the MSNBC host, taking to the Senate floor the next morning to school the young Senator from Kentucky on the how-tos of millennial outreach: "If Mr. Paul wants to be taken seriously," the seventy-seven-year-old senator said, "he needs to do more than pull political stunts that fire up impressionable libertarian kids in their college dorms."[26] Senator McCain's most reliable sidekick, Lindsey Graham, sided squarely with Barack Obama and his stonewalling attorney general. "I do not believe that question deserves an answer," Graham said. According to a *Washington Times* report:

> *Mr. Graham said he defends Mr. Paul's right to ask questions and seek answers, but said the filibuster has actually pushed him to now support Mr. Brennan. Mr. Graham said he had been inclined to oppose the nomination because he'd found Brennan to be qualified for the job but also "arrogant, kind of a bit shifty." He said he wasn't going to filibuster him but would have voted against him on final passage, but now he'll vote for him. "I am going to vote for Brennan now because it's become a referendum on the drone program," he said.*[27]

But while the filibuster was actually going down, most ignored it. It was just a stunt, another archaic parliamentary procedure that no one really pays any attention to. Early on during Paul's almost thirteen-hour talkathon, the story was not a story at all, but a source of ridicule among Beltway cognoscenti. So few rational people outside the Capitol Bubble pay any attention to what happens on the floor of the U.S. Senate on any given day, you could almost forgive the denizens of conventional wisdom for missing the point. Why would an outsider like Rand Paul, who won his Senate seat in Kentucky in 2010 by beating Republican leader Mitch McConnell's handpicked candidate in the Republican primary, use such an insider tactic?

DEFENDING THE STATUS QUO

The filibuster, a last-ditch attempt by a single member of the Senate to stall consideration of legislation, has a storied history in legislative warfare. Typically, this roadblock has been used to defend the status quo inside the cloistered walls of the most closed, insulated institution in America—the U.S. Senate. They don't call it a club for nothing: It's a privileged cadre unaccustomed to the bright light of public attention. And that's the way they like it.

The most infamous use of the filibuster, of course, was by Democratic senator Strom Thurmond, then a segregationist who famously fought against the efforts of Martin Luther King. In 1948, Thurmond had actually left the Democratic Party to run for president as a Dixiecrat. Thurmond would later argue that "King demeans his race and retards the advancement of his people." [28]

In 1954, the landmark Supreme Court case of *Brown v. Board of Education* had ended "separate but equal" and started the process of integrating schools all over the country. A year after that, Rosa Parks famously refused to sit in the "blacks only" section of a bus in Montgomery, Alabama.

Her bravery helped force racist government policies into the public psyche.

On August 28, 1957, at 8:54 P.M., Thurmond took the Senate floor in opposition to major provisions of the 1957 Civil Rights Act. He would not stop until more than twenty-four hours later. He denied that any blacks were being denied a right to vote and argued that every state already had sufficient voter rights protections in their existing laws. "I think it is indicative that Negroes are voting in large numbers. Of course, they are not so well qualified to vote as are the white people." [29]

To this day, Thurmond's remains the longest verbal filibuster in U.S. history. Ultimately, the Civil Rights Act of 1957 passed the Senate and was signed into law, but not before Thurmond and his Democratic colleagues had stripped the legislation of key provisions. [30]

Segregationists like Thurmond had thoroughly corrupted the notion of "states' rights" and the Tenth Amendment to the Constitution—a vital and legitimate check on federal abuses of power—to obfuscate their real agenda. Thurmond and many others used the excuse of federalism to justify the oppression of individuals—unequal treatment under law—but that was never the intention of the federalist system. Yes, the states must not submit to federal tyranny, but that does not give them license to be tyrannical themselves. It was all about *the rights of the individual*.

Free people should judge others based on the content of their character, not the color of their skin. If you believe in liberty and the dignity of the individual, you inherently believe in treating everyone equally under the laws of the land. This is a first principle. It's nonnegotiable. Defending the rights of the individual, including equal treatment under the law, is a fundamental responsibility of a constitutionally limited government, as James Madison had so eloquently argued in *Federalist* 51.

#STANDWITHRAND

So the filibuster seemed like a strange tactic for a feisty young "tea party" senator to employ against presidential tyranny, and in support of the fundamental right of individual American citizens to due process under our constitution. He was flipping old traditions upside down, using an old tactic to disrupt the status quo, using the same tool that Strom Thurmond had used, *but this time to defend our individual civil liberties from a government that had overstepped its constitutional limits.*

The Internet had changed everything. As Rand Paul spoke, social media, especially Twitter, exploded almost instantaneously as people tuned into C-SPAN based on a tweet or a Facebook post from a friend. *Check this out: This senator is speaking truth to power.* At their peak, creative Twitter hashtags like #StandWithRand, #filiblizzard, and the more prosaic #RandPaul were seeing two thousand mentions a minute.[31] A report put together by the media analysis firm, TrendPo, shows that, by the filibuster's end, #StandWithRand had become the number-one trending topic on Twitter with more than 1.1 million associated tweets. During the following twenty-four hours, Rand Paul gained an astonishing 44,700 new followers.[32]

When Rand Paul finally yielded the floor, hoarse and exhausted, it was past midnight on March 7. He had been speaking for a total of twelve hours and fifty-two minutes.[33] Mocked and ignored just a few hours earlier, Paul had succeeded in changing the conversation globally, transforming the political landscape by endrunning the Beltway information monopolists.

That day, a *Washington Post* headline pronounced: "An Old Tactic Packs New Power in Digital Age." Polling of possible 2016 Republican presidential contenders showed Paul jumping the queue "into Tier 1 status, leapfrogging Paul Ryan, Jeb Bush, Chris Christie, and Mike Huckabee to place a very respectable 2nd in a 9 person field."[34] On March 8, one CNN commentator said that, "[a]t least for the time being, Tea Party darling Sen. Rand Paul is the effective leader of the Republican Party. And that's a pretty big deal."[35]

Was #StandWithRand a publicity stunt, a waste of time? I don't think so. Public opinion polls conducted in the days and weeks following the filibuster showed a marked decline in approval for the Obama administration's drone policies. Whereas a clear majority of Americans favored drone strikes prior to Paul's raising of the issue, afterward approval for such policies plummeted to just 41 percent.[36] On March 7, Attorney General Eric Holder sent another letter. "Dear Senator Paul," it read. "It has come to my attention that you now have asked an additional question.

'Does the president have the authority to use a weaponized drone to kill an American not engaged in combat on U.S. soil?' The answer to that is no."[37] This, of course, was the question all along. Opinions will come and go, and polls will rise and fall, but ultimately, Senator Paul's efforts produced real policy change, a reining in of executive power, and a legal opinion from the top law enforcement officer of the land, previously not offered, that will have legal standing in future debates over our civil liberties and the limits on executive power.

A RIGHT TO KNOW

Nearly 215 million people use Twitter on a regular basis, sending half a billion tweets a day. Facebook has more than a billion active users worldwide.[38] Twitter is scooping old media on news stories approximately 20 percent of the time.[39] More than 50 percent of people say they have learned about breaking news from social media rather than from a traditional news source; 27.8 percent of people get their news from social media in general. Social media was responsible for breaking major news stories such as the Egyptian uprising, the Hudson River plane crash, and Osama bin Laden's death.[40] Information and knowledge itself have been democratized, no longer filtered by three television networks offering a monolithic product controlled by self-anointed experts.

The full implications of tremendously responsive and ever-mutating social networks are only beginning to be understood, but their decentralizing power is undeniable. Freedom is trending, I believe, because of the "long tail" of the Internet. Imagine a world where ideas are easy to find, and learning is available to anyone who's willing to work for it.

John Perry Barlow, the cyberlibertarian and cofounder of EFF, saw the disruptive nature of the Internet before just about anyone else, and imagines a world where every one of us has, at our fingertips, a "right to know":

> For the first time, we have it within range to make it possible for anybody, anywhere to know everything that he is intellectually capable of assimilating about any topic. That is to say, he can know as much, whether he is in the uplands of Mali or midtown Manhattan, about some nuance of molecular biology as is presently known by anybody. Now, I understand that knowledge also has a context, and this is much likelier to be the case if it's not in the uplands of Mali, since you won't have so many people to discuss it with. But there is, I think, the possibility that we can convey to future generations the right to know. The right to know, as much as they want to know, and that includes everything that's presently known and generally applicable by anybody. And it also includes everything that is known, or can be known, about what one's government is doing.[41]

Several days after Rand Paul's game-changing talkathon, I happened to be speaking to the annual gathering of the European Students for Liberty, in Leuven, Belgium. My topic, fortuitously, was all about the liberating nature of social media and the strategic implications for grassroots efforts to restore liberty and the dignity of the individual in an era of encroaching government power. What amazed me most about the gathering was the incredible growth in the number and quality of the students in the group from the previous year's event. There was now a packed auditorium, more than four hundred freedom-loving students gathered from all over the world. Given the timeliness of the Paul filibuster, I had decided to use it as a real time example of the new political disintermediation. How many people knew of it, I asked. Every one of their hands went up. How many people participated in the Paul filibuster on Twitter? Almost every hand went up.

How quickly things have changed. Imagine this opportunity before us. We have the ability, at least theoretically, to find every single person in the world who believes in individual freedom and who has access to the Internet. We can connect with them, share ideas, books, and strategies. We can gather and coalesce, build a virtual division of labor, and generate a new accountability against the many instances of government overreach and tyranny happening every day, all over the world. Together, acting in voluntary cooperation, we can create a "greater social intelligence" and social awareness unlike anything possible before.

Compare the experience of these students, many of whom had already read the same books I struggled to find when I was their age.

I stumbled upon the ideas of liberty by accident when I bought that Rush album, *2112,* the one I didn't want. I happened upon an old used copy of *Anthem* at a community garage sale. I accidentally discovered the Austrian economics program available at Grove City College and George Mason University in a late night argument fueled only by the wisdom-enhancing properties of cold beer.

Today, I would just Google it. I would "like" Ludwig von Mises on Facebook. I would watch a 1976 performance of "2112" on YouTube. I would follow Rand Paul, or Ted Cruz, or Justin Amash on Twitter. I would almost instantaneously connect with centuries of intellectual tradition that had allowed subsequent generations to stand on the intellectual shoulders of the best activists, entrepreneurs, thinkers, rabble-rousers, and disrupters.

Imagine a world where Lech Walesa, the heroic grassroots leader of Poland's Solidarity movement, could have live-streamed his courageous calls for civil disobedience from the shipyards of Gdansk. What if Samuel Adams, America's first community organizer, could have live-tweeted the Boston Tea Party in real time across the colonies? What if Dr. King could have organized a virtual March on Washington for all of those who could not afford to get to the nation's capital on August 28, 1963?

The Internet is a force multiplier for free people because we naturally fit with its ethos. Everything is transparent, and there are simple rules. No one gets to tell anyone else what to do. But people are constantly coming together in common purpose, based upon mutually agreed-upon goals, to bigger ends. That is precisely how freedom works.

That's why freedom is trending online. Freedom is breaking down barriers to knowing. It also seems to be breaking down the old rules of political partisanship. It may no longer be so black and white, Republicans versus Democrats. It may not even be about liberals versus conservatives. What if the new political spectrum has on one side those people who want to be left alone, those who want to be free, those who don't hurt people or take their stuff, and on the other extreme of this new scale stands anyone who wants to use government power to tell you how to live your life?

Don't believe me? Consider this story from the *Guardian* about a grassroots protest in Washington, D.C., organized against the Obama administration's practice of mass surveillance of its innocent citizens:

> *Billed by organizers as "the largest rally yet to protest mass surveillance," Stop Watching Us was sponsored by an unusually broad coalition of left- and right-wing groups, including everything from the American Civil Liberties Union, the Green Party, Color of Change and Daily Kos to the Libertarian Party, Freedom-Works and Young Americans for Liberty.*
>
> *William Evans, of Richmond, Virginia, may have best embodied the nonpartisan atmosphere and message of the event. He wore a "Richmond Tea Party" baseball cap, as well as a Code Pink sticker saying "Make Out, Not War." He is a member of the Richmond Tea Party but not of Code Pink, he said, adding that he "just loved" what the sticker said. Evans said he was attending to protest the "shredding of the constitution" and added that he was happy that "you guys on the left are finally starting to see it."*

"We may not always agree on our belief system," he added, *"but thank God we agree on the Constitution."*[42]

Code Pink and the tea party? The Internet really does change everything. No wonder John McCain is freaking out.

CHAPTER 7

A SEAT AT THE TABLE

THE OLD WAY OF doing things in Washington was based on a closed system, an exclusive club that favored insiders and the politically connected over principled leaders with big ideas. Follow the leader. Toe the party line. Shake the right hands. That was the only way to get elected, the only way to even have a shot at making a difference. But within the constraints of the system, the rules are always stacked against freedom, and accountability, and fiscal responsibility.

The emergence of the Internet and social media has begun to change all that. The machinery of government no longer functions entirely behind closed doors, shielded from the light of public attention. Information on last-minute floor votes and arcane congressional floor procedure is tweeted out, posted, and otherwise instantaneously distributed to millions of concerned citizens. Through the magic of live streaming, we can watch events unfold on the House and Senate floors in real time, from the comfort of our own smart phones.

Knowledge is power, and the diminishing marginal costs of getting good information about Washington's ways is changing the old, tired political calculus. Politicians can no longer hide from their constituents, telling them one thing back home while voting for business as usual in the nation's capital. As a result, we are beginning to see real accountability, and the effects, though only just beginning to be felt, are amazing.

Thanks to the power of political disintermediation, the American people are making their voices heard in Washington. A new generation of congressmen and senators has emerged to give voice to the formerly voiceless, to keep their promises, and to stand on principle.

This is nothing short of a paradigm shift that gives shareholders a real seat at the table in Washington. Our proxy representation at the board of directors' table is a growing bicameral "Liberty Caucus," the size and quality of which is historically unprecedented in American politics.

I was lucky enough to sit down with six of the most exciting figures to emerge from this new political environment, to get their take on things. I asked them about their history with the ideas of liberty and their experiences confronting the political establishment. Senators Rand Paul (R-KY), Mike Lee (R-UT), and Ted Cruz (R-TX), along with Representatives Justin Amash (R-MI), Thomas Massie (R-KY), and David Schweikert (R-AZ), are leading among those writing the new rules in politics, where power does not go to those most entrenched in a broken system, but remains with the people, where the founders intended it.*

These are some smart, fearless guys. Because this is my book, I took the liberty to mash up six separate conversations into an imaginary gab fest between all six legislators. All of their quotes, of course, are the real thing.

Here's what went down in my imaginary living room:

MK: *We're talking about the ideas of liberty and the way that the world has changed so much in the last couple of years, but I wanted to ask you first, how you got into these ideas. How did you discover freedom?*

TED CRUZ: As a kid I got very involved in a group in Houston that was called the Free Enterprise Institute. It had a program where it taught high school kids principles of free market economics, and it would have us read Milton Friedman, and Hayek, and Von Mises, and Bastiat, and have us prepare speeches on free market economics. In the course of four years of high school I ended up giving right about eighty speeches across the state of Texas on free market economics, and also on the Constitution. And that became really the intellectual inspiration and foundation for being involved in the liberty movement.

---

* The interviews in this chapter were conducted separately, with responses edited together to form a simulated "round table" discussion. I have made every effort to preserve the context of my guests' remarks so as not to misrepresent them in any way. The full, unedited interviews were videotaped and will be available for viewing on donthurtpeople.com.

DAVID SCHWEIKERT: It came to me as a teenager. Somehow I got my hands on an Ayn Rand book. And unlike most people, I started with a book called *We the Living*. In Arizona it's really hot during the summer, so you're just inside going through the pages. And I fell in love with the heroine in that. And from there it just sort of built into understanding the power of the individual. And I have to admit, even in the high school I was in, there were probably a dozen of us who became Rand devotees.

THOMAS MASSIE: My gateway issue was gun rights. When I was eighteen I went to school in Massachusetts from Kentucky. I'd read about people who wanted to ban guns, but I'd never met one. And instantly I found myself surrounded by these people that wanted to ban guns. And that was my liberty issue.

MIKE LEE: I was raised with a real love of the country. My parents taught me that America is a special place, that America is unlike other countries. And we're very privileged, we're very fortunate to live here because of these shared values and the heritage that we have inherited from prior generations. My parents taught me about the structure and how it's set up from an early age. One of the things that I've been frustrated with since at least the age of ten is the fact that the federal government is doing too many things. We were always supposed to have a limited-purpose national government, a federal government with only a few basic responsibilities. It was supposed to perform those really well, and it was supposed to take care of those things to the exclusion, in many circumstances, of state authority. But outside of those areas, it was supposed to stay out and let state and local governments take care of the rest, along with civil society.

JUSTIN AMASH: I was done with college, done with law school, and noticed that my views on politics were a little bit different than some of my Republican colleagues. It was the [George W.] Bush era of Republican politics. So, I decided I'd do a Google search and threw some of the terms into Google that I thought matched my viewpoints. Up popped F. A. Hayek.

I like Hayek's style. It's an intellectual style. There's a strong focus on spontaneous order, the idea that order pops out of our free interactions with each other. I found that very appealing and when I read Hayek's works, they really struck a chord with me.

MK: *Hayek talks about how individuals come together in voluntary association and create institutions, and those institutions both inform, and are a constraint on our behavior. I always thought that that interplay between community and the individual made a lot of sense and explains how the world holds together and works so well without some benevolent despot telling us what to do.*

JUSTIN AMASH: Yeah, that's absolutely right. He's very good at making the distinction between government and society. There can be societies where people interact, where they cooperate, where they form groups together. But you don't have to have a government deciding how all of those interactions work.

MK: *The left loves to use this atomistic caricature that we're all Ayn Randoids, selfish individuals willing to do absolutely anything to get what we want. But that's the complete opposite of what I get out of Rand. Her work was really focused on individual responsibility. We need to take the word "community organizer" back, I think, and take the word "community" back.*

JUSTIN AMASH: Right, and a lot of what libertarians are about this idea that people work together, that they cooperate, that they form these sort of social groups. That's perfectly acceptable as long as they're voluntary associations.

MK: *Let's talk about politics for a little bit. I think that we're in the midst of a realignment, maybe even a paradigm shift. That same disintermediation, decentralization, more power to the individual dynamic is happening in our politics. And people like you are beating establishment candidates with all of the traditional advantages: more money, more people jetting in from Washington, D.C., to endorse them. Tell me that story.*

THOMAS MASSIE: I think the old model was that you ran for state legislature and you became a state representative, then you became a state senator, and you were a

good party player, a good team player, and then somebody recommended that you get into a congressional
race, and you come up the ranks. That's been turned
on its head.

There are some guys here in Congress that have
never held an elective office. Ted Yoho, he's great. He's
a large-animal veterinarian. Jim Bridenstine, he's great.
He was a Navy fighter pilot. Neither of those guys held
an elective office, and they beat an incumbent Republican in a primary to get here to Washington, D.C. That's
only possible with grassroots support. Social media is
part of it. Alternative media through talk radio is part of
it. It's enabled a different model of coming to Congress.
You have the grassroots, these outside organizations
like FreedomWorks, which are immensely important in
the races, and not just in the races, but after the race
is won in influencing these congressmen when they get
here.

TED CRUZ: I think there is a fundamental paradigm shift
happening in the political world across this country,
and that paradigm shift is the rise of the grass roots. In
the Texas Senate race, when we started I was literally at
2 percent in the polls. Nobody in the state thought we
had a prayer. My opponent was the sitting lieutenant
governor, who was independently wealthy. He ran over
$35 million dollars in nasty attack ads against us. And
what we saw was just breathtaking. We saw first dozens, and hundreds and then thousands and then tens
of thousands of men and women all across the state be-

gin rising up, begin knocking on doors, begin making phone calls, and going on Facebook, and going on Twitter, and reaching out and saying, "We can't keep doing what we're doing. We are bankrupting this country. We are threatening the future of the next generations if we keep going down this road."

It was breathtaking, the grassroots tsunami we saw. Despite being outspent three to one, we went from 2 percent to not just winning, but winning the primary by fourteen points and winning the general by sixty points. It was an incredible testament to the power of the grass roots, and I think that's happening all over the country.

DAVID SCHWEIKERT: I've had a handful of brutal political elections. It feels like every time I run I end up having the establishment folks against me, because I'm not sure that they want some of their little special deals examined or taken away. And what you're finding is that the activists, the public, because of that access of information through the Internet, are sort of learning, "Oh, this is reality. This is my alternative, and there are options that do work."

THOMAS MASSIE: We look at communist countries and socialist countries and see how the Internet has changed them, or the countries that are led by despots. When they get the Internet, they sort of start coming around and there are revolutions there. That is happening here, we just don't notice it. But it's happening slower, because we've got a peaceful process for doing that.

RAND PAUL: Our Facebook [following] is now bigger than several of the news networks'. I'm not saying that to brag, I'm saying that because there is power in Facebook. There is power in Instagram. There's power throughout the Internet. It really has led to an amazing democracy.

MIKE LEE: And I think it's important to point out, Matt, that that is not our power. That is power that we have from the people. It is power that we have only because we connect to the people, and only to the extent that we connect to the people. What has changed is that, with the power of social media and other new channels of communication, the *Washington Post,* the *New York Times,* and the small handful of media outlets that have in the past been the exclusive conduits of information about what's happening in Washington no longer have a monopoly. The cartel is broken, and with the breaking of that cartel, the people are empowered. And they're empowered by a new generation of elected officials who are there to stand for the people and not for their own perpetual reelection, and not for the perpetual expansion of government. That's a game changer. That's how we bring about the restoration of constitutional government.

MK: Do you think something different is going on in terms of the relationship of Americans with their federal government?

RAND PAUL: Yeah, and I think the people are probably ten years in advance of the legislature, and probably always are. The grass roots and the public react in a way, but it takes a while for their will to get transmitted to Washington. Why? Because incumbents win almost every race around here. So there are people who were elected in 1980. They're still representing the people in 1980 who first elected them. Each successive election becomes easier, and they're not listening as carefully to the American people. So, the new people, we're listening pretty carefully. We just got elected.

THOMAS MASSIE: Most congressmen come here with the best of intentions. They want to do the right thing. But eventually they're like zombies. They get bitten and they become part of the zombie mob, and they vote with their party regardless of what's in the bill. Some people can make it a month without getting bitten, and some people can make it a whole term. But eventually, just like a bad zombie series on TV . . .

DAVID SCHWEIKERT: Past scandals were often about an individual engaging in a bad act. Now the public is understanding that there is this collective movement of bad acts. And it's about the preservation of power. The only way to break that down is to radically change those institutions or completely eliminate those institutions and move to a very different model.

RAND PAUL: And I disagree with some people who say we're too conservative or too much in favor of balanced budgets or too much for lower taxes and less regulation. No, we can be all of those things. We don't need to lose what we're for, but we also have to be for a bigger message of liberty. Young people don't have any money. You ask young people about regulations and taxes and they're like, "I don't have any money. I don't own a business. But I've got a cell phone and I'm on the computer, and I don't like the government snooping on whether I read *Reason* magazine or whether I go to FreedomWorks' website. I don't want the government to know that unless I'm accused of a crime." They care about privacy, but they may not care about taxes. So, we don't give up on taxes, but we also need to talk about issues that young people are interested in.

DAVID SCHWEIKERT: There's incredible opportunity, particularly with that under-forty, under-thirty-five population. We have data that says they're brand switchers. When they walk into the grocery store they don't buy Tide because their mom and grandma bought it. They buy what they think is the best value, or what they saw on their social media as having a benefit they want. And I think, actually, they're about to grow up politically. They have to now realize that they've been lied to by this president about privacy—look at the things the NSA has done—about their individual freedoms. This White House has not cared about their individual freedoms. As a matter of fact it's been more collectivist. I'm waiting for that revolution with young people to say, "And now you've basically made me an indentured servant through the debt, through my future tax liability, and now what you've done to me health-care-wise." It's time for our young people to wake up and understand: The battle's on.

MK: It's an interesting time to be here right now, because we're in the midst of this gargantuan fight. Not just for the soul of the Republican Party, but perhaps for the future of this country. What does this new party look like?

RAND PAUL: It looks like the rest of America if we want to win. I say, "With ties and without ties, with tattoos and without tattoos." It needs to look like the rest of America, but also in an ethnic way as well. We are a very diverse culture. We need to reach out to African-Americans and say, "Look, the war on drugs has disproportionately hurt the black community." One in three black Americans is a convicted felon, primarily because of nonviolent drug crimes. We need to reach out and say, "It isn't fair that we're targeting black Americans for arrest." It is said, by surveys, that whites and blacks use drugs at about the same rate, and yet the ACLU recently said that blacks were being arrested at five to six times the rate. Prison statistics show that seventy-five percent of prisoners are African-American or Latino, and it is because the war on drugs is not equally applied. I think we need to tell kids that drugs are a bad thing. I tell my kids to stay away from drugs. They're a bad thing. But if one of my kids gets caught, I don't want them in jail forever. I saw the other day, Michael Douglas's kid is in jail for ten years. He's been in solitary confinement for two years.

Is he hurting himself by using drugs? Absolutely. But I would rather see him in some kind of rehabilitation hospital than solitary confinement.

But we have to understand, and as Republicans we need to go to the African-American community and say, "Look, they're losing not only their freedom. They come out and then they're a convicted felon for the rest of their life." You ever try to get a job? They call it "checking the box." Checking the box of convicted felon. They can never get ahead again. Their child support payments build up while they're in prison. They come out and they owe four thousand dollars in child support. How do they ever pay that working minimum wage, or not working at all? One thing adds up and it's this cycle of poverty. I think if Republicans had a message, that message is a limited-government one. This is: The government should protect us from violence against other individuals. The sort of self-inflicted bad things that people can do to themselves, we should try to work as a society to minimize that, but putting people in jail for doing bad things to themselves is just not good for society.

JUSTIN AMASH: And when I go back to my district my constituents are very supportive of what I'm doing, Republicans and Democrats. I think things are changing, and I talk to many of my colleagues who are just entering Congress the last two cycles. They think more like I do on many of these issues. And in fact, when you look at the NSA amendment, for example, newer members were much more supportive of my amendment than members who have been here for a long time. I think there's a generational shift and it's shifting in the direction of libertarianism.

TED CRUZ: I think the Republican Party needs to get back to the principles we should have been standing for in the first place. We need to get back to defending free market principles and defending the Constitution. I think what we're seeing, the rise of the grass roots, is the American people holding elected officials accountable of both parties. I think that's terrific. I think that should happen a lot more.

MK: But if you were to open to a page in the New York Times, *they would describe a libertarian as socially moderate and fiscally conservative. I never thought that was quite right. I always thought it was about our relationship with the government and whether or not we got to control our own lives.*

JUSTIN AMASH: That's right. It's just about being able to make decisions for your own lives. So, there are very socially conservative libertarians. I'm a fairly socially conservative libertarian. And there are other libertarians who are not as socially conservative. But the idea is that we should have a government that allows us to make those decisions for our own lives, and we can decide as a society whether we like those values or not. And if you disagree with someone, you're free to tell them. But we don't need government imposing one viewpoint on everyone.

THOMAS MASSIE: People like to label everybody in Washington, D.C. I've been called a libertarian-leaning Republican, a constitutional conservative, a tea party congressman, but I think the one that fits best is when they call me one of the twelve members of the Republican conference who didn't vote for John Boehner.

MK: It strikes me that it's no longer so much about Republicans versus Democrats. It may be about D.C. insiders versus the rest of America.

JUSTIN AMASH: Yeah, I don't think we should ever worry about who we're working with in terms of Republican versus Democrat versus libertarian or independent. We have to work together here.

TED CRUZ: There is a divide, and it's a much bigger divide than a divide between Republicans and Democrats. That's the divide between entrenched politicians in Washington and the American people. There are a lot of people in both parties in Washington who just aren't listening to the people anymore.

DAVID SCHWEIKERT: Well, think about this. If you're a bureaucrat, what do we know about bureaucracies, of every kind both private and public? Ultimately the preservation of the bureaucracy becomes the number-one goal.

So, if you're Lois Lerner, you're at the IRS, is it as much even ideological, as it's the preservation of the bureaucracy? And you see that all over Washington, because the scandal at the IRS isn't the only place this type of activity is taking place. It's up and down government. Because are you going to support the party, the more collectivist party that wants to grow government, wants to give you bonuses, wants to give you certain shiny objects, or the party that wants to hand power back to the states? You end up with a very different incentive system, and it's quite perverse.

THOMAS MASSIE: Here's one thing that people don't really understand, that I didn't understand. They say that money corrupts the process. I've always kind of believed that, but I didn't know how it corrupts the process. Congressmen raise a lot of money. Some of them raise two or three million dollars an election cycle. But what do they do with that money? Their reelection is virtually assured. It's more certain than anything that they're going to come back, because they're incumbents and they've done the right things. So, they don't need the money to get reelected. They're not buying yachts and Ferraris with the money. What are they doing with the money? It's the currency of power.

Here's what they're doing with the money that they raise: They're giving it to other congressmen. And then they become ingratiated. They feel like they owe that congressman something. A vote, maybe on an issue, or a cosponsorship on a bill. So, it's the currency of influence within Congress, and then you also take most of that money and you give it to your party, whether you're a Republican or a Democrat.

There's a big football game that's going on in Congress. The party that has the majority is playing hard to keep the majority, and the party that's in the minority is playing hard to get the majority. It's a football game that's played with money, and the more money you raise for your party, the more influence you're granted. You'll get a better committee. Do they measure committees in terms of how much you can do for your constituents? No. All the A-committees in Congress are based on how much money will lobbyists give to you if you get on one of those committees. So, you raise money for your party, you're a good soldier, you get on a higher fund-raising-capable committee, you raise more money, but now you've got a quota. Now you're on the treadmill. If they give you a spot on the big pirate ship, you've got to collect a lot of treasure. That's the way the process gets distorted. That's how money distorts the process in Washington, D.C.

RAND PAUL: Imagine how it could be if Hillary Clinton is the nominee for the Democrat Party. If she's the nominee and she wants to be involved in the middle of the Syrian civil war, and she doesn't give a damn about your privacy. Imagine if, on the Republican side, we have someone who wants a constitutional foreign policy, who says, "Sure, we defend our country. You mess with us, you're going to get what happened after 9/11—

overwhelming use of force against you. But we're not going to be involved in every civil war, and Congress will vote. The will of the people will decide whether we're in war." I think you could have a complete transformative election, where all of a sudden the reactionary, nonthinking individual is going to be Hillary Clinton, and the Republicans could have a forward-looking person who talks about privacy and talks about adding a degree of justice to our criminal justice system.

MK: *The newly empowered citizenry, with their new tools of accountability, makes me an optimist, even though everything in this town and everything that President Obama has done to our economy and to our Constitution should make us despair about the future. Are you an optimist or are you a pessimist about the future of this country?*

THOMAS MASSIE: This place is way more broken than I realized before I came here. Now that I'm here, I give it a fifty percent chance that we're going to be able to turn this ship before we hit the shore. And a fifty percent chance that it's going to take something big to wake people up and to get the changes we need. But the only thing I can do is fight to turn the ship. That's what I'm working on. Instead of being home in Kentucky and preparing for the ship to hit the shore, I'm up here trying to avoid the shore.

MIKE LEE: I'm an optimist in a Churchillian sort of way. Winston Churchill is reported to have said, "The Amer-

ican people can always be counted on to do the right
thing, after they have exhausted every other alternative." I think we're reaching that point where we have
exhausted every other alternative, and we will be left
with doing the right thing. That's what the American
people are doing. That's what they're saying. They want
to return to a time when the people are sovereign, and
they're citizens, not subjects.

TED CRUZ: I'm incredibly optimistic. I'm optimistic because I think there is a movement that's sweeping this
nation of millions of Americans who are waking up and
looking around. If you look at the past year, the rise of
the grass roots, in fight, after fight, after fight in Washington, the grass roots have turned the fight around.
Nothing scares elected officials more than hearing from
their constituents, and in my view, liberty is never safer
than when politicians are terrified.

DAVID SCHWEIKERT: I'm optimistic also, but be careful because sometimes I'm pathologically optimistic. How do
you get the public, mom and dad, the young person, the
person who's trying to grow their life and their business,
to be able to take that little bit of their time? And it's
not about writing a check, though those are helpful. It's
about reaching out to a FreedomWorks or other organizations and driving their voice, saying, "We're paying
attention, and we care."

CHAPTER 8

TWELVE STEPS

WHAT, EXACTLY, *DO YOU want*?

I get this question all the time, inside the Capitol Beltway. Sometimes the hostility of the inquiry makes me feel like I'm participating in the drug intervention of an old friend. You've finally got their attention, and they feel trapped, busted. Then comes the denial, the paranoia, and the hostility. An addict will shoot at any messenger that delivers the bad news: You have a big problem, and the path you have chosen will not—cannot—end well.

This is precisely the way that official Washington has reacted to the citizens asking the tough questions of their two party representation. Obviously, those who ask this question typically have an agenda. They are trying to deflect attention, boldly claiming that Washington does not have a spending problem. An addiction to power? Not here; at least nothing that can't be solved by giving the fixers another fix, more and more money and control. Without another government program, how will anything get better? The relentless clamor for more of your money rattles through Washington like junkies pleading for just one more hit.

One of the common critiques coming from progressives, the media, and chin-clutching establishmentarians inside the Beltway is that we are just against things. President Obama loves this particular straw man. We oppose a government takeover of health care, so we must be against people getting health insurance. We oppose federal meddling in education, so we must be against children learning. We oppose an omnipotent surveillance state, so we must be against the safety of innocents.

David Brooks, the resident "conservative" at the *New York Times,* doesn't even try to hide his disdain for the new generation of legislators who have come to Washington committed to changing the rules of the game:

> *Ted Cruz, the senator from Canada through Texas, is basically not a legislator in the normal sense, doesn't have an idea that he's going to Congress to create coalitions, make alliances, and he is going to pass a lot of legislation. He's going in more as a media protest person. And a lot of the House Republicans are in the same mode. They're not normal members of Congress. They're not legislators. They want to stop things. And so they're just being—they just want to obstruct.*[1]

Harry Reid went so far as to call us "anarchists," simply because we oppose funding an expensive federal health-care takeover that the president himself has arbitrarily repealed or delayed in part some twenty times so far.[2] The senator most responsible for drafting the legislation, Democrat Max Baucus, called it "a huge train wreck coming down" in April 2013.[3] But now we are the "anarchists" for insisting that the government not fund, with borrowed money, something that no one in D.C. seems to think will actually work. They are acting like desperate addicts, aren't they?

How do we get from here to there, to more freedom and prosperity? How do we get from where we are today—with ever more encroaching government control, unimaginable fiscal liabilities, and so few in Washington, D.C., willing to do what needs to be done—to the point where the federal government is back to its limited and proper role?

Public choice economists might tell us that it's impossible, that governments naturally, inexorably, march forward—like the White Walkers descending on Westeros in *Game of Thrones*—expanding to the point where they choke off productive initiative, and great nations die. Think Rome, and the tragic devolution from a republic to an autocratic empire, and then to the dustbin of history.

How can we reverse course and make sure that America doesn't go down that fateful path of no return? To me, this is the most interesting strategic question that constitutional conservatives and small-*l* libertarians—moms and dads who just want a better life for their kids—have to answer.

The solution will never be a quixotic fix of more "revenue" or another top-down reorganization of your life by some faceless bureaucrat who knows nothing of you and your family and doesn't much care. We need a better, more compelling freedom agenda. The burden on us will always be far higher to explain how freedom works.

We understand our principles. We get freedom. We know that simple rules of personal conduct—Don't Hurt People and Don't Take Their Stuff—create tremendous upward potential for all of us, and that opportunity for all creates peaceful cooperation. Even though the insiders tell us the opposite, we know that open societies actually spread the wealth, and that closed, top-down systems lock in the spoils of the Haves at the expense of generations of Have-Nots. We understand the ethos of liberty that is ingrained in every one of us makes America an exceptional place.

So what, exactly, should we do to restore liberty?

This chapter lays out a twelve-step policy agenda: positive, innovative ideas that would improve people's lives by letting them be free, by spending less of your hard-earned money on someone else's favors, by letting you choose, by treating us all equally under the laws of the land.

Radical stuff, I know.

1. COMPLY WITH THE LAWS YOU PASS

As Steve Forbes likes to say, the planners in Washington should have to eat their "own cooking." This seems like such common sense, but you won't be surprised to learn just how controversial this idea is behind the closed doors where congressional staffers and career bureaucrats congregate. Do as I say, they prefer, not as I do.

Former Obama administration Treasury secretary Timothy Geithner, who was confirmed by the U.S. Senate to enforce your compliance with complex federal tax laws, didn't even see fit to pay his own taxes,[4] apparently believing himself above such prosaic responsibilities.

Back in 2011, it was revealed that House Democratic leader Nancy Pelosi and other key members of Congress and their committee staff had played the market with the inside information of what their proposed laws would do to the stock valuations of certain industries.[5]

This sort of behavior is emblematic of the contempt shown by Congress for the laws they impose on the rest of us. While the STOCK Act[6] purported to put an end to congressional insider trading, the substance of the legislation was later rolled back before being implemented, by unanimous voice vote. Members of the House were not given time to review the bill that Senate majority leader Harry Reid had sent over in the middle of the night.

"Rather than craft narrow exemptions, or even delay implementation until proper protections could be created, the Senate decided instead to exclude legislative and executive staffers from the online disclosure requirements" of the STOCK Act, reports the Sunlight Foundation.[7] So the bicameral vote that insisted that D.C. insiders comply with the same trading laws as the rest of us was public and virtually unanimous, but the gutting of the law carries few legislators' names or fingerprints.

More egregious still are the constant attempts by members, staff, and federal employees to exempt themselves from ObamaCare. House Ways and Means Committee chairman Dave Camp wants to change that, offering a proposal that would place all federal employees, even the president himself, into the same exchanges required by the rest of the country.

"If the ObamaCare exchanges are good enough for the hardworking Americans and small businesses the law claims to help, then they should be good enough for the president, vice president, Congress, and federal employees," Camp's spokeswoman explained.[8]

2. STOP SPENDING MONEY WE DON'T HAVE

American families have to balance their budgets. The government should do the same. This is not rocket science.

Why is it so hard for Congress to balance the budget? The core problem, of course, is that they are not spending their own money. They are spending your money. The ghost of John Maynard Keynes provides them with a pseudo-intellectual rationale to "stimulate aggregate demand." But we are on to them and know that the only real stimulus they are buying with borrowed money is for their own reelection prospects.

Given that, as of this writing, the national debt tops $17 trillion, it seems like common sense would dictate a few things:

- Stop new spending on new programs.
- Prioritize dollars and get rid of programs that don't make the cut as top priorities in a world of scarcity.

- No sacred cows allowed until we solve the problem, so put everything on the table.
- Deal honestly with entitlements by acknowledging unfunded future promises.
- You can't tax your way to a balanced budget without tanking the job creation that actually generates tax receipts.

I know, more radicalism. Harry Reid is so offended by these budget principles that if you agree with them, he thinks you are an "anarchist."

So many in both parties have grown comfortable simply kicking the can down the road and rubber-stamping an endless series of increases in the "debt ceiling," or short-term "continuing resolutions" that claim deficit reduction in future years while spending more today. But it's really not that hard to map out a plan to clean up Washington's fiscal train wreck. In fact, FreedomWorks "crowdsourced" ideas for a citizens' "Debt Commission" that would balance the budget in just a few years. Senator Mike Lee tried to bring those ideas to his Senate colleagues in November 2011 and was literally evicted from the Russell Senate Office Building by staffers representing Senators Chuck Schumer (D-NY) and Lamar Alexander (R-TN).[9]

No, this isn't an *Onion* spoof. I'm not making that up.

Senator Lee has introduced a constitutional amendment that would require Congress to balance the budget each year and limit spending to 18 percent of GDP, the forty-year average of federal receipts.[10] It was the basis for a consensus balanced-budget amendment that the entire Senate Republican caucus eventually signed on to.

The Congressional Budget Office has released a report suggesting that if nothing is done to control spending, by 2038 the federal debt could be as high as 190 percent of GDP.[11] At that point we can send congressional emissaries to Athens, Greece, to solicit innovative budget savings ideas from the Hellenic Parliament.

3. Scrap the Tax Code

The federal tax code should only exist to fund the necessary functions of government.

Special interests and congressional deal making have corrupted the tax code beyond anything imaginable in 1913, when Congress passed the Sixteenth Amendment to the Constitution, authorizing a national income tax. This incomprehensible complexity favors insiders and the special provisions they lobbied for, and the rest of us foot the bill. It's political class warfare against working Americans. The problem isn't tax cuts for the rich; it's a tax code that prevents working Americans from getting rich.

Complexity also enriches bureaucratic advantage. Complexity means more career public employees to navigate ambiguous rules. The tax code becomes a weapon in the hands of IRS agents who have a partisan or parochial agenda, or hold a grudge.

We need to scrap the code, and abolish the IRS. We

need to clean out the whole building, hose it out, and start over with a simple, low, flat tax. The government function of revenue collection should be limited and straightforward. No agendas, no social engineering, no overbearing discretionary authority in the hands of gray-suited soviets.

Senator Ted Cruz (R-TX) has proposed doing exactly that. "We ought to abolish the IRS and instead move to a simple flat tax, where the average American can fill out our taxes on a postcard," Cruz told Fox News. "It ought to be just a simple, one-page postcard and take the agents, the bureaucracy out of Washington. And limit the power of government."[12]

The most powerful case for tax reform is a moral one, the common cause of blind justice. If you don't trust Washington, D.C., to give you a fair shake, why not just treat everyone equally under the laws of the land?

Making the tax code simple, low, fair, and honest would be a powerful means of unleashing human potential. Class warriors on the left would howl about the injustice of treating everyone equally, but their real agenda is in defending the Beltway interests that have designed the current mess.

The true victims of fundamental tax reform are the insiders who have carved out their favors, as well as the legislators and bureaucrats who make their living off soliciting, creating, and navigating new complexity. The reduction in wasted time and money devoted to compliance would unleash capital, job creation, and upward mobility, while the elimination of complex loopholes would level the playing field between Americans and tax compliance enforcers inside government.

4. PUT PATIENTS IN CHARGE

Okay, so we all agree that ObamaCare is exactly the wrong medicine. We need to repeal the whole thing and start over. That does not mean that there is nothing wrong. But the answer is in more freedom, not the coercive hand of government bureaucrats.

The system as it exists today is plagued by a lack of competition and by complex labyrinths that prevent patients from taking charge of their own care and treatments.

The singular problem with our health-care system is all of the faceless, gray-suited middlemen standing between you and your doctor. So-called "third party payers" are the direct result of government distortions in health-care markets. Remember, allowing employers to provide benefits like health care, with pretax dollars, was a political fix to FDR's wage and price controls.

What if we cut out the bureaucrats, and their take, and let you make the choices right for you and your family? Would providers work harder to satisfy your needs? Would you get more quality at a lower price?

Of course you would.

There is a simple way to free patients and doctors from third parties like employers, HMOs, the IRS, or the faceless deciders at HHS. This could be accomplished by eliminating the punitive bias in the tax code that taxes health insurance and services when purchased directly by individuals. This would be a pretty simple fix that empowers patients without some complex, top-down redesign by the federal government. If health care is different, and vitally important to all of us, let's provide care for our families with our own hard-earned dollars, before the federal government takes its cut. In other words, treat everyone the same, regardless of where you work and whom you work for.

Other commonsense reforms include health savings accounts for younger workers, stripping all of the "mandated benefits" from gold-plated insurance plans that drive up both costs and overconsumption of health services. We should also let families shop for better health insurance policies in all fifty states, just like any other product we might shop around for.

No mandates, no coercion. Just choice, and providers who work for your health and your return business. Politicians like to make empty promises about "universal coverage," even though they can't possibly provide for it. Besides, the goal should be *better health care at lower costs,* and Washington is particularly ill-suited to provide *that.*

Health care is a fundamentally personal issue. The relationship between a patient and doctor needs to be based on trust and mutual understanding. Let's stop robbing patients of their privacy, their dignity, and their freedom to choose.

It's really not that complicated, unless health-care reform is more about their control over you than it is about your control over your health care.

5. CHOICE, NOT CONSCRIPTION

One-size-fits-all entitlements take security and control away from you, and that's exactly upside down. You should have a say and a choice in your own future plans.

Of course, it's hard to talk about health-care reform without talking about Medicare and Medicaid. Both programs are in dire financial condition. It makes no sense to take $500 billion out of Medicare, as ObamaCare does, to spend on the creation of a new program. It also makes no sense to expand Medicaid, as ObamaCare does, to grow Medicaid populations and financial obligations that are already bankrupting state budgets.

The biggest challenge with the federal budget is the so-called entitlement programs like Social Security, Medicare, and now ObamaCare. Already, these programs consume a big part of the total federal budget. And this ominous trend does not even take into account what the Trustees Report for Social Security and Medicare estimates are unfunded promises in excess of $100 trillion.[13] The total fiscal gap of all our government liabilities is $222 trillion, according to economist Laurence Kotlikoff.[14] I know, it's almost impossible to wrap your mind around that one.

Such programs will literally consume the entire budget if we don't rethink this forced, one-size-fits-all approach to questions like providing seniors' health care and retirement benefits.

Giving people choices is the key. Today, so much is forced, mandated, and controlled by someone else. If these programs are good and desirable, we should let people choose. After all, choice and competition are the fundamental building blocks of customer satisfaction.

I don't think we should change the rules of the game on retirees and near retirees already locked into the current system. That would be wrong. But so is forcing young people into one-size-fits-all programs that experts do not believe will be around when future retirees arrive. So young people should be free to choose. It would be wrong to force them into a system they can't count on in retirement.

Besides, we no longer work for the same company our entire lives, like our grandparents did. New systems need to be mobile and stick with us, under our control.

Senator Rand Paul's budget proposal for fiscal 2014, "A Clear Vision to Revitalize America," recognized that entitlement programs are insolvent and on track to bankrupt the nation. He proposed, among other things, replacing involuntary enrollment with individual choice, allowing young people to opt out of Social Security if they think they can get a better deal elsewhere.[15]

Opting out costs the Social Security system money today, but it also takes unfunded future liabilities off the books. Honest accounting would demonstrate the value of young people taking more personal responsibility. People work hard for their money. It is only reasonable to let them choose how to use it to invest in their own futures, especially when the fiscal health of the entire nation hangs in the balance.

And we know what Congress has done to the Social Security Trust Fund. They have already spent all of the retirement funds on other stuff. Really.

Medicare is also a major source of conscription into the federal benefits programs. If seniors want to receive the Social Security that they have paid for their whole life, they must also enroll in a bloated government health insurance program that suffers from a lack of competition on the open market. Shouldn't seniors be allowed to choose their own health care, rather than being forced into a system they may not like, want, or need?

We should make participation in Medicare voluntary. Why not let seniors choose for themselves? If you don't want to participate in Medicare, you shouldn't have to. The system could use the money. We should also let participants in Medicare purchase additional health-care services outside the government system, and let doctors provide those services without being penalized.

In 2012, Senator Paul introduced the "Congressional Health Care for Seniors Act," a bill that would have allowed seniors to sign up for the same health insurance program enjoyed by members of Congress, the Federal Employees Health Benefits Program.[16] Unlike Medicare, this would open up competition and allow seniors more choices over their health care. There are currently 2,250 participating plans in the FEHBP, so there would certainly be no shortage of options. Furthermore, it's estimated that this plan would save more than $1 trillion in the next decade.

As John Kerry once said in his endorsement of a similar program back in 2004, "If it's good enough for us, it's good enough for every American."[17]

6. END INSIDER BAILOUTS

The bridesmaid of big government is always some well-heeled interest that wants a special deal. If the government weren't so involved, insiders would have to go back to serving consumers and taking responsibility for their own actions.

Some call it crony capitalism, but I think that gives honest entrepreneurs a bad name, smeared by the corrupt behavior of beggar CEOs seeking new handouts. One of the biggest problems in Washington, D.C., is the unholy collusion between favor-seeking "businessmen," committee chairmen, and White House operatives. Can't meet consumer demand? Can't compete with smaller, more prudently run banks? Don't know how to turn a profit on "green" technology? Get on your G5 and jet to Washington. There, someone will make you an offer you can't refuse.

Why is it that powerful Wall Street banks and multinational car manufacturers get bailouts paid for by the rest of us?

Well, who's your man in D.C.? What's the name of the well-heeled lobbyist in charge of getting you special favors and goodies from government? Don't have one, do you? And there's the problem. The trend in D.C. is toward more consolidation, more "insider trading," where favored interests—think General Electric or Solyndra or the government employees' union or the city of Detroit—rearrange the rules and federal budget allocations to their advantage.

As long as the favors are being handed out, someone

other than you, someone with insider pull, is going to get in line first.

The best weapon against this insider cronyism is transparency, public shaming, and market accountability. But there are some innovative ideas to deal with too-big-to-fail investment banks and the other trough-feeding interests that grow fat on your tab.

The market dominance of unaccountable investment banks has been fed by a de facto understanding that bad behavior will be bailed out. You will never find that statute, but when the crisis comes, the irresponsible risk takers will hold us hostage, and the political class will fall in line. Recall Nancy Pelosi's final plea for votes to pass the TARP bailout:

> *It just comes down to one simple thing. They have described a precipice. We are on the brink of doing something that might pull us back from that precipice. I think we have a responsibility. We have worked in a bipartisan way.*[18]

Don't doubt that the insiders in D.C. will find common ground with Wall Street's bad actors when it matters. Huge special interests are protected at the taxpayers' expense, even when they display gross incompetence and an inability to act responsibly. We saw it with TARP, and again with the General Motors bailout.

If you believe in freedom, you understand that future rewards entail risk, a willingness to put your money where your mouth is. This, to me, is the cool part of capitalism; it allows everyone to play in the rough-and-tumble scrum of serving consumers better. Maybe you have a better idea, or see efficiencies no one else does.

But if you get it wrong, freedom holds you to account. No looking to someone else to bail you out. The same should be true if the "you" is named Citibank or AIG or Countrywide Financial. If bad behavior isn't allowed to be corrected by the relentless accountability markets, bad actors will double down on risky behavior, creating a politically generated boom-and-bust cycle with no end.

As chairman of the House Committee on Financial Services, Jeb Hensarling (R-TX) seems like the odd man out in his lonely fight to unwind Fannie Mae and Freddie Mac. The committee has traditionally protected the cozy—and highly profitable—relationship between big banks and these so-called "government sponsored enterprises." That's Washington-speak for the socialized risk that taxpayers bear, and the personalized profits for certain insiders with the right political pull.

"The two largest, most influence-exerting, regulation-avoiding, bailed-out institutions weren't banks and weren't located on Wall Street. They were Fannie Mae and Freddie Mac, the mortgage market financial Frankensteins that were created not in a competitive marketplace, but in a government lab in Washington," Hensarling said.

In 2011, he introduced the "GSE Bailout Elimination and Taxpayer Protection Act," a bill designed to stop the ridiculous taxpayer-funded payouts to the government-sponsored enterprises (GSEs), Fannie Mae and Freddie Mac, both of which contributed in no small measure to the housing crisis of 2007. "The GSEs are on track to be the nation's biggest bailout, more than AIG and GM and all the big banks combined. It's time to enact fundamental reform of Fannie and Freddie before these companies go from 'too big to fail' to 'too late to fix.' "[19]

Americans should get a fair shake, with equal treatment under the law, rather than being forced to prop up failing enterprises with their tax dollars. Free markets are all about accountability, and that means both profit and loss. That's the American way.

7. LET PARENTS DECIDE

Parents know the educational needs of their children best.

Every day we are told that America is falling behind in educational standards, that we are in danger of being unable to compete on a global scale, that our children aren't learning well enough, fast enough. The proposed solutions invariably include putting more good money into a bad system, tightening the grip of the federal bureaucracy on education standards, lengthening school hours, imposing more rigorous testing, and separating children ever further from their parents' control, putting them into the hands of Big Brother.

These two trends are the inverse of each other: The more top-down control from Washington, the worse our kids perform in monopoly schools. This seems like another one of those "teachable moments," doesn't it?

From "No Child Left Behind" to "Common Core," all of these top-down, one-size-fits-all federal programs seek to deprive parents of options in the way their children are schooled. Common Core standards eliminate choice at every level, hobbling states, localities, teachers, students, and parents in their ability to choose education standards that work for children, over those that are arbitrarily mandated from on high.

All of this is in direct conflict with the empirical evidence that children learn best when parents are free to choose from a variety of educational options to suit the individual needs of their children.

Education belongs at the local level. Only parents in local communities are well equipped to decide the policies that work for their kids. The fight against Common Core standards is largely being waged at the state level, but in order to reform education in the long run, we need to get gray-suited bureaucrats from faraway Washington out of the business of managing your child's education. Do they know what your son needs? Could they possibly care more than you do? Do they even know your daughter's name?

The U.S. Department of Education does nothing but stand in the way of preserving choice and keeping education local, where it should be. I think we should shut it down and put tax dollars back in the hands of parents and allow them to choose the right school for their children, be it public, private, charter, or home.

A freedom-based education policy puts parents first, recognizing that they are the ones best placed to choose what is right for their own families. This is common sense, knowing that personal knowledge guides the individual talents of your kids.

8. RESPECT MY PRIVACY

In our constitutional system, one of the sacred laws of justice is "innocent until proven guilty." We are supposed to be protected against unreasonable search and seizure, and law-abiding citizens should have a reasonable expectation of privacy from the all-seeing eyes of government surveillance.

The exploits of the NSA reveal that such constitutional protections are under attack, and that the Obama administration has little regard for the Fourth and Fifth Amendments to the Constitution. The "guilty until proven innocent" philosophy of government-by-surveillance is a fundamental perversion of the American principles of justice.

Americans should be free to live their lives without the fear of government constantly snooping into their every activity. We do not want a police state in which we are watched at all times, with the powers that be waiting eagerly for any opportunity to inflict punishment to keep us in line.

Freedom is compromised when surveillance is pervasive. Treat people like criminals and you will make criminals out of them. The activities of the NSA should be reserved for actual lawbreakers, always conducted under the rule of law with properly issued warrants.

Gradually, we have allowed our privacy rights to slip away, starting with the warrantless wiretaps of the Patriot Act and extending to the outrageous domestic spying program of the NSA. A freedom-based policy would restore privacy to the American people and reassert the principle of "innocent until proven guilty."

Justin Amash has taken the lead in attempting to end NSA spying on Americans once and for all. Winning the bipartisan support of an impressive array of congressmen, his "USA FREEDOM Act" offers bold new ideas to respect the privacy of ordinary citizens and check the power of government spying.

"The days of unfettered spying on the American people are numbered," said Amash. "This is the bill the public has been waiting for. We now have legislation that ceases the government's unconstitutional surveillance. I am confident that Americans and their representatives will rally behind it." [20]

The bill is a multi-pronged attack on the surveillance state. It ends the collection of Americans' data by the NSA except in cases of suspected criminal activity; it requires FISA court decisions to be made available to Congress, and summaries of those opinions to be released to the public; it gives telecommunications companies more freedom to disclose information on government surveillance to the public; and it installs a special advocate to argue in favor of preserving Americans' civil liberties before the FISA court.

9. END THE FED MONOPOLY

Monopolies don't work very well when it comes to maintaining high quality and a low price. It's the lack of accountability and competition that leads to expensive, inferior outputs. This seems like a good analogy to explain why the Federal Reserve has trashed the dollar. A lack of accountability and competition has degraded your purchasing power. A dollar's just not "as good as gold" anymore.

This is one of the major problems with the Fed. It's both "independent," yet systematically manipulated by political insiders. Congress made things worse in 1977 when it amended the Federal Reserve Act to create a so-called dual mandate, which amounted to a blank check for the Fed to do just about whatever it wants. "The Board of Governors of the Federal Reserve System and the Federal Open Market Committee shall maintain long run growth of the monetary and credit aggregates commensurate with the economy's long run potential to increase production," the mandate states, "so as to promote effectively the goals of maximum employment, stable prices and moderate long-term interest rates."

If you wonder what exactly this means, you're not alone. The only thing that is certain is that the ambiguity of "maximum employment" gives incredible discretion to Fed micromanagers operating in an incredibly secretive fashion. With a lack of supervision or oversight, there is no way of knowing what really goes on within its deified walls. The first step in eliminating the Fed's monopoly is a comprehensive audit to find out exactly in which ways it has been mismanaging our currency, and how it is managing the $3.7 trillion in toxic assets from mortgage-backed securities it has acquired in recent years.

To the extent there is a central bank, its only job should

be to protect the integrity of the currency, not to manipulate the dollar based on pressures from politicians and big investment banks. The Fed would be more accountable and predictable if it operated using rule-based monetary policy rather than the blank check discretionary power it has today. We should eliminate "maximum employment" from the Fed's current dual mandate. Economists can't even agree on what full employment actually is, let alone understand the infinitely complex price signals that drive market decisions. Giving the Fed a mandate to do whatever it wants leads to irresponsible abuses of the currency and drives the political business cycle of boom and bust.

Should we end the Fed outright? Should we adopt a gold standard that prevents the easy manipulation and expansion of a "paper" currency? I think we start by "denationalizing" money, an idea first proposed by F. A. Hayek. Let's legalize gold and other electronic payment systems as a means of exchange. Let's allow competition in currency. Choice, transparency, and competition would end the Fed monopoly, and stop the destructive boom and bust of monetary manipulation.

A monetary policy consistent with freedom rests firmly on the idea of sound money, free from manipulation by insiders, bureaucrats, and politicians. "Freedom of our currency is the fundamental issue," wrote my college economics professor Hans Sennholz. "It is the keystone of a free society."

10. Avoid Entangling Alliances

Congress, more closely accountable to the people, should approve acts of war. Wars cost precious American lives, and will always drain our economy of resources.

Remember George Washington's caution not to "entangle our peace and prosperity in the toils" of other nations' affairs. He was worried about the security of Americans first, and he knew that the budget implications of foreign entanglements mattered in a very real way. It is not isolationist, as some neo-conservatives accuse libertarians of being. It's about opportunity costs, economic realities, and common sense. "As a very important source of strength and security, cherish public credit," Washington counseled. "One method of preserving it is to use it as sparingly as possible, avoiding occasions of expense by cultivating peace."

In September 2013, President Obama seemed ready to go to war with Syria. The situation was a complex one, with rebel forces no more sympathetic to American interests than the incumbent regime. This echoed the situations in Egypt and Libya several years earlier, where American intervention clearly did not improve things either for our own interests, or that of those countries' own citizens.

Recognizing that both Congress and the American people were overwhelmingly against the idea of intervention in Syria, the president verbally toyed with the idea of acting without congressional approval, in violation of the War Powers Act.

Fortunately, it didn't come to that, but this was not an isolated incident. There has been a recent string of overseas military operations conducted without a formal declaration of war from Congress, ever since the War on Terror blurred the lines between enemy combatants and common criminals.

This is a dangerous precedent. James Madison once wrote in a letter to Thomas Jefferson: "The constitution supposes, what the History of all Governments demonstrates, that the Executive is the branch of power most interested in war, and most prone to it. It has accordingly with studied care vested the question of war to the Legislature." Madison rightly recognized that the power to send soldiers to their deaths and drop bombs on other nations should not be vested in one man alone.

The American people knew better. Our national security depends on our economic strength and fiscal stability, and it would be reckless for Congress to bankrupt us in the process of becoming the world's policeman in someone else's civil war.

Accountability and a new restraint on executive branch power came from the people. As *National Review* put it: "The Phone Lines Melt," referring to an unprecedented grass roots onslaught of opposition to the president's proposed war. "And inboxes are inundated, as people urge their congressmen to oppose military action in Syria."[21] Pretending that military measures won't cost more than D.C. "experts" predict ignores *everything* conservatives already know about experts' predictions. It doesn't matter what government program we're talking about—whether we're debating Social Security, Medicaid, or therapeutic air strikes in the Levant. Costs always exceed illusory budget baselines.

There is an opportunity cost to war. Resources directed toward building tanks and bombs cannot be used for more productive purposes. As every freshman economics textbook once held, if you want more guns you have to give up some butter.

Admiral Mike Mullen, former chairman of the Joint Chiefs of Staff under President Obama, has argued that America's debt is the single largest threat to our national security. The economic strength of our nation is the basis of our leadership in the world. Mullen says:

> [T]he most significant threat to our national security is our national debt. . . . That's why it's so important that the economy move in the right direction, because the strength and the support and the resources that our military uses are directly related to the health of our economy over time.[22]

A foreign policy based on the ideals of freedom would address these problems. The use of military force is a serious thing not to be employed lightly. It is a mistake, therefore, to get involved in entangling alliances that force our troops to act when doing so is not in the national interest, when there is no clear objective or definition of victory, and when the lives of innocent civilians would be unnecessarily forfeit as a result.

11. DON'T TAKE PEOPLE'S STUFF

The right to be secure in your property is the cornerstone of a free society. The founders knew this, and for most Americans this is a commonsense proposition that keeps the government from arbitrarily taking their stuff. Yet today private property rights are being threatened by an expanding and unresponsive government. More and more citizens are finding their property under attack, either through a growing web of onerous regulation, or outright seizure through aggressive use of eminent domain and civil forfeiture laws.

Property rights were a core issue for the thirteen colonies chafing under British rule. Thomas Jefferson, James Madison, and Alexander Hamilton wrote extensively on the importance of private property, and asserted that you didn't need the government to grant you rights in your property; property rights precede government and are inherent to the rights of all individuals. This view helped frame the U.S. Constitution and its constraints on government power. In fact, six of the ten amendments in the Bill of Rights touch on the question of property.

The most explicit protection of private property in the U.S. Constitution is the Fifth Amendment, including the famous "takings" clause, which states, "No person shall be . . . deprived of life, liberty or property, without due process of law; nor shall property be taken for public use without just compensation." The founders saw the government's potential to expropriate property and drafted the Constitution to limit this possibility.

Yet as government grew, the protections of private property enshrined in the Constitution were weakened by legal decisions and the growing scope of the regulatory state. Today, numerous government actions threaten the private property of individuals, whether through excessive regulation, expanding government control of our nation's resources, or abuses of the legal system that take our property rights.

Eminent domain abuse has been on the rise, with many individuals losing their property as local governments seize it in the name of economic development. The developers win, but homeowners often get the short end of the stick. As of late 2013, the Institute for Justice was suing seven U.S. cities over their attempts to seize private property, but the issue first came to popular attention when the Supreme Court upheld the *Kelo v. City of New London* decision in 2005, which allowed city officials in New London, Connecticut, to seize homes and businesses on the pure assertion that new development would provide jobs and new revenue for the city.

The decision stacked the deck in favor of business interests at the expense of small property owners. As noted in Justice Sandra Day O'Connor's dissent to the court's opinion, "Any property may now be taken for the benefit of another private party, but the fallout from this decision will not be random. The beneficiaries are likely to be those citizens with disproportionate influence and power in the political process, including large corporations and development firms. As for the victims, the government now has license to transfer property from those with fewer resources to those with more. The Founders cannot have intended this perverse result."[23]

The release of the *Kelo* decision created a surge of public outcry against governments taking private property. Unfortunately, the outrage has dissipated with time, and eminent domain abuse continues to cause problems. While some states have attempted to address the issue, strong federal legislation would help us all sleep more soundly. Passing the Private Property Rights Protection Act would be an important step forward.

Perhaps more disconcerting is the rise in civil forfeitures, where law enforcement and other agencies seize property from criminal enterprises. More and more innocent people are being caught in an ill-defined dragnet that has stripped them of their property. Consider the recent case of a grocer in Michigan. Despite receiving a clean bill of health from the IRS less than a year before, the local family business had its bank account wiped out in January 2013 after a secret warrant was issued over deposit transactions that allegedly violated banking laws. In fact, there were no violations, and the activity in question was due to the need to comply with the store's insurance policy. Yet that grocer in Michigan is still waiting for a day in court to plead his case.[24]

The vast increase in data collection by the federal government and the rise of the government's Big Data policies are putting more and more people under a spotlight. Unfortunately, they may never know, because many of these decisions are made by faceless bureaucrats with warrants granted by secret courts. Civil forfeiture laws should be wiped from the books; without being convicted of a crime, no individual should have to hand over his private property to the government.

Besides outright takings of property, the growing regulatory state that we live in also threatens our property and livelihoods. The federal government has stopped people from building homes in the name of environmental protection; local governments are passing new laws to keep food trucks from competing against local restaurants, alternatives to taxis, such as Uber, are being threatened with regulation prompted by taxicabs, and the IRS is now deciding to issue new regulations about who can be a tax preparer. At every turn the growth of regulation is a threat to our property and our liberty. And more often than not, it is wielded by those with political clout against those without. The regulatory state offers political insiders more levers to press and more avenues of access in order to protect their interests from new businesses trying to enter the market. Paring back the regulatory state—which costs the nation $1.2 trillion a year—is a sure way to enhance the freedoms we enjoy.

12. Defend Your Right to Know

The Internet has changed everything. Creating a digital community that spans the globe has led to unprecedented disintermediation as individuals gained the freedom to interconnect on their own, no filters, no hierarchy, no middleman required. EBay, the online auction house, has made every individual a potential retailer, while allowing customers an unprecedented scope of access to retailers across the world. Access to information and the ability to communicate across all the corners of the world have empowered individuals in ways that were inconceivable even a few years ago.

Why do you think it is that tyrants of all stripes now go after control of the Internet and readily available social networking platforms first? They want more control, and political disintermediation online shifts control and freedom—and a real voice—to the end user. "Most of the world's dictators share a common fear," argues Joel Brinkley, a Pulitzer Prize–winning journalist now at Stanford University. It's social media. "Facebook, Twitter, blogs and the rest have spread around the world and are now being used as cudgels against authoritarian leaders in places like Vietnam, Russia, Belarus and Bahrain. In those states and so many others, the leaders are attacking tweeters and bloggers as if they were armed revolutionaries." In Iran, "bloggers are given long prison terms or sentenced to death, charged with 'enmity against God' and subverting national security."[25]

The implications are profound, particularly in terms of participatory politics. In the fight for freedom, the Internet is everything, and we should fight to protect it from government encroachment and censorship.

As more and more of our lives are carried out online, the data cloud is growing, and so is the potential for abuse. The government can now readily access private information, as when the IRS illegally seized 60 million personal medical records.[26] At the same time, the rules governing federal access to online information are murky as to whether a search warrant is required. As Declan McCullagh noted, "An IRS 2009 Search Warrant Handbook obtained by the American Civil Liberties Union argues that 'emails and other transmissions generally lose their reasonable expectation of privacy and thus their Fourth Amendment protection once they have been sent from an individual's computer.' "[27]

It is important, then, to ensure that the liberties enshrined in the Constitution extend to every sphere of activity—the Constitution does not stop where technology begins.

Many of the current laws governing online privacy were written for a world that no longer exists. For example, the Electronic Communications Privacy Act (ECPA)—which sets the rules for law enforcement agencies accessing private data online—was written before anyone heard of Facebook or Dropbox. Online storage was expensive, and no one envisioned a world of cloud computing; data was only protected from warrantless searches for 180 days, because no one could possibly store information any longer than that. Consequently, under ECPA, any data older than 180 days are fair game for law enforcement. No warrants are necessary.

As is often the case, technology evolved in ways that the lawmakers in Washington could not envision. Today, virtually all Internet users engage in some form of cloud computing, whether it's Facebook, an online music collection, or simply archiving emails. As a result, much of our lives is accessible to law enforcement agencies without ever needing a warrant. The laws must change with the times. Senators Mike Lee (R-UT) and Pat Leahy (D-VT) have offered an amendment to ECPA to make it clear that government must obtain a warrant prior to accessing private online information.

In addition to arbitrary incursions into individual privacy, a growing government presence on the Internet poses significant threats to free speech and online activism.

Both government policy and businesses seeking refuge from the intense competition of the Internet may introduce barriers that ultimately limit consumer choice or access to information.

In 2011, this drama played out in Washington, as Big Hollywood and other content providers sought tough new laws to stop Internet infringements on their material. In the Senate, the debate focused on PIPA—the Protect Intellectual Property Act. In the House the debate targeted SOPA—the Stop Online Piracy Act.

Intellectual property has long been the topic of heated debates because the definitions are not clear and the exceptions ambiguous. The founders understood the need to balance innovation with intellectual property. Article I, Section 8 of the Constitution—often called the copyright clause—states that Congress has the authority "To promote the Progress of Science and useful Arts, by securing for limited Times to Authors and Inventors the exclusive Right to their respective Writings and Discoveries."

A period of exclusive ownership or copyright provides an incentive to produce works that might otherwise not be undertaken. At the same time, this unique clause suggests that the founders viewed intellectual property differently from other forms of property, so much so that it is addressed separately.

Since first establishing a copyright of fourteen years in 1790, the span of protection has increased dramatically, thanks to pressure from interested parties. Today, it stands at the life of the author plus another seventy years, or in the case of corporate authorship, 125 years from the creation or ninety-five years from the year of publication, whichever comes first.[28] These politically defined "rights" seem like a subversion of the founders' intent.

How much more inspiration to innovate does a 125-year copyright provide?

This largesse to powerful business interests has always been balanced by the doctrine of fair use, which, under certain circumstances, allows limited use of copyrighted materials without first seeking permission from the owner of the copyright.

The Internet poses a new threat to intellectual property owners, allowing individuals to copy and transmit content, often at almost zero cost. SOPA and PIPA were pushed by Hollywood interests in response, to clamp down on piracy. But these ill-conceived measures effectively set up the infrastructure for the federal government to censor the Internet, granting unprecedented authority to shut down millions of websites that failed to meet the new standards. In effect, these bills would have made the government the official online enforcer, mandating search engines and third parties to remove links to websites deemed unacceptable.

While these efforts to censor the Internet were defeated by a broad coalition of grassroots and civil liberties organizations, new threats loom. A new, more sweeping proposal is CISPA, which stands for the Cyber Intelligence Sharing and Protection Act. This legislation would provide broad new powers to the government. It would allow "companies to identify and obtain 'threat information' by looking at your private information," according to the Electronic Frontier Foundation. "It is written so broadly that it allows companies to hand over large swaths of personal information to the government with no judicial oversight—effectively creating a 'cybersecurity' loophole in all existing privacy laws." [29]

Restrictions on the flow of information have important political implications. Regulation and other formal constraints on the Internet have the potential to shape the information available to individuals and therefore the political debate. We can't go back to the world of three nightly news channels and have the same level of political discourse that we do today. The Internet has to remain free from government control and unnecessary regulation, free to provide activists a platform to educate and mobilize, and free to anyone wishing to exercise their First Amendment rights to free speech.

ALL THESE ISSUES COULD be acted upon by Congress this year, if the political will were there.

If the will were there. How many times have you heard that before, all the while watching our "representation" in Washington drive headlong over the cliff? The lemmings seem utterly unaware, or at least wholly unconcerned, that a dire end quickly approaches. And that's the point. They won't do the right thing when left alone; they will run our country right off the edge, pointing the finger of blame at one another even as they plummet to their own undoing.

The fact is that government control has become a narcotic for D.C. power mongers. One hit, and most get hooked, scrambling for more, lashing out at those who would deny them another. Legislators and executive branch kleptocrats lack the will to act because they simply can't make it without their next fix. Even when the desire for change is there, the compulsion to spend is simply too overpowering to resist. Lawmakers can't break the habit on their own. They mindlessly consume new tax dollars, and fake printed dollars and even dollars borrowed from China, like zombies on the hunt for fresh brains.

They need help. It's time for you to intervene.

This twelve-step program is designed to wean the government off the empty promises of new entitlements, excessive spending, and unchecked executive power. It seems utterly crazy to keep doing what we did before, to follow the old rules of bipartisan collusion, if doing so does not solve problems. We need to scrap the tax code, and balance the budget and restore respect for the simple rules embodied in our Constitution that treat everyone just like everyone else.

We can do all these things if and when America beats Washington. That's the key. The perfectly constructed constitutional amendment or the best patient-centered health-care reform goes exactly nowhere if Washington is left to its own devices. You will have to act.

CHAPTER 9

NOT A ONE-NIGHT STAND

*The most dangerous man to any government is the
man who is able to think things out for himself,
without regard to the prevailing superstitions and
taboos. Almost inevitably he comes to the conclusion
that the government he lives under is dishonest, insane
and intolerable, and so, if he is romantic, he tries to
change it. And even if he is not romantic personally he
is very apt to spread discontent among those who are.*

—H. L. MENCKEN[1]

DO YOU EVER FEEL like politicians want just one thing from
you? That maybe, just maybe, they don't really care about
you, your dignity, or your freedoms at all?

To be sure, the political courtship can be awesome.
There's always lots of sweet talk. Politicos know all the right
buttons to push, always telling you exactly what you want
to hear. They call you. They write to you. They send you
notes in the mail. They "friend" you on Facebook. Some-

times you get a personal text message on your cell phone, or even an invite to hang out with George Clooney. How cool is that?

Democrats seem more comfortable courting you online, or at your front door. Republicans typically prefer expensive grand gestures, like a national, thousand-point-saturation television ad buy. The GOP is old-school that way.

They promise you a transparent, honest government. Would you like to see a simple flat tax that doesn't have all of the carve-outs and special deals for others in it? How about a balanced budget that stops stealing from future generations of your family? Do you want more choice and control for your own retirement, or the freedom to determine your own child's education, or even to defend your right, as a patient, to choose your own doctor?

They even pledge to keep their promises, and to stay faithful the day after.

It might be worth taking a chance, you think.

You know they only want one thing, one time, on the first Tuesday in November. You *know* they are not looking for a long-term relationship, that their fidelity to principle will suddenly disappear when they get back to Washington, D.C. But the charm offensive wears down your defenses. The letters and the calls and the posts and tweets and the thirty-second spots and the big promises are just too tempting. You want to believe it, because the future of your country, and your children's future, is at stake. Sooner or later, you cast aside your inhibitions, and you do it.

You vote for the same guys that let you down last time. And it *never* works out.

DUMPED, AGAIN

I'm not judging here. I've done it, too. I stand in line to vote (in the District of Columbia, no less). I have written checks to candidates for public office. I have hoped for the best. I have even walked precincts, door to door, for someone else's preferred candidate, who's running on someone else's bad ideas, all because they promised me they would do the right thing.

I always wake up, the day after the election, feeling used. Used again. It never works out. They never call the next day. They don't write. They don't text. And they never, ever keep their long-term commitments.

That's the problem with political parties. The relationship always turns out to be a one-night stand that leaves you feeling used, ignored, and then dumped for someone or something that's far more attractive, someone or something waiting back in Washington, D.C.

Consider the sorry state of President Obama's signature health-care legislation circa January 2010. It was jammed through Congress using parliamentary trickery because the people of the very blue Massachusetts decided to send a clear political message in the special election of Republican Scott Brown. "We don't want this," Bay Staters said at the ballot box. "We don't trust Washington to oversee a massively complex redesign of our health-care system." No matter the will of the people. Nancy Pelosi used "deem and pass" procedures so that the Senate would not have to provide the sixty votes that Senate majority leader Harry Reid no longer had.

Senator Arlen Specter of Pennsylvania, who had switched from Republican to Democrat in hopes of clinging to political power, had provided the deciding vote for the Senate bill. Specter, of course, would not have been a senator except for the extraordinary efforts of President George W. Bush, Karl Rove, and the GOP establishment to protect him in his 2004 primary challenge from Pat Toomey.

Come hell or high water, the establishment was going to do good for themselves regardless of earlier promises to stay true to you.

And then there were the many "read my lips" promises from President Barack Obama, always intended as lies to the American people to provide political cover for those Democrats jamming through sweeping, unread legislation that no one wanted. He promised greater transparency and efficiency. He promised an end to the cronyism that always attends a major rewrite of the rules of the game. He promised that you could keep the health insurance you had if you liked it. He promised that his new plan would not ration care. He promised that your family's health-care costs would go down, not up.

Everyone knew he would break these promises. He just wanted one thing: your vote, in 2012.

The Obama White House has arbitrarily delayed or repealed various provisions of the Affordable Care Act without consulting Congress, even though the legislative branch of our government has the sole responsibility of passing, and repealing or amending, the law. The executive branch is supposed to enforce the law. Under Article II, Section 3 of the Constitution of the United States, the president "shall take care that the Laws be faithfully executed." Except when he doesn't want to? Only when it's politically advisable? I'm no constitutional lawyer, but I don't see the wiggle room here. This utterly outrageous and arbitrary process of a president choosing to only implement the parts of laws he likes seems so un-American, even autocratic.

But never mind; no one seems willing to stand up to him. He's the president, they say. He has the bully pulpit. We will fight, after the next election, they promise, hoping we don't notice the inconvenient truth that after 2014 comes 2016—another election and another excuse not to stand up.

Dumped again, you hopeless romantic.

To be sure, there are a few brave souls who have stood up, the growing minority in Washington that can be counted on. I'm referring to members of Congress with a seat at the table who ran for office in 2010 and 2012 on the solemn promise, if elected, to do everything in their power to replace ObamaCare with policies that respect patients, not bureaucrats. True, most politicians run on promises to respect your civil liberties, to be prudent with the spending of your cash, or to be deferential to your rights to determine your own health-care choices. But these new guys seem to actually mean it.

And this is a crisis. All of the experts, and political operatives, and the octogenarian pooh-bahs who opine from the Senate floor, and "unnamed sources" from congressional leadership staff and "senior officials," unleashed a united brick wall of hate and venom and "expert advice" against those that would do, in Washington, D.C., what they promised to do back home when soliciting your trust and your vote.

They accuse us of creating false hope among conservatives and libertarians and Tea Partiers and independents who just want to be left alone that we can actually win this fight.

Maybe they just don't want us to fight at all?

THE HOMOGENIZING PROCESS

Beltway dinosaurs, Democrats and Republicans, are uniting against a new generation of leadership in Washington, leaders like Mike Lee and Ted Cruz and Justin Amash and Thomas Massie and Rand Paul and David Schweikert, all of those who seem so willing to challenge the old way of doing things.

The *National Journal* asked some of the old bulls what they think about this new type of legislator. Former Republican House Speaker Denny Hastert, for one, pines for the days when campaign finance rules gave party bosses all the power. "The people you got usually weren't too far to the left or to the right. The party was sort of a homogenizing process," he says.[2]

According to *Merriam-Webster*, to homogenize is to make something uniform or similar. To standardize, unite, merge, fuse, integrate, or amalgamate. Make it the same as all the rest. So the party bosses who got us into the fix we are in want to go back to the way things were before, to do things the same way that they were done in the past. One of Hastert's former deputies in the House of Representatives agrees, arguing that newly empowered grassroots organizations are too disruptive. "FreedomWorks is not serving the legislative process well by telling these old guys to just buzz off." Now, Hastert frets, candidates "have to worry constantly about primary challenges."

What are they really worried about? A little competition? Accountability?

According to Peter Schweizer, a research fellow at Stanford University's Hoover Institution, politics was a lucrative profession for the former House Speaker:

> *When Hastert first went to Congress he was a man of relatively modest means. He had a 104-acre farm in Shipman, Illinois, worth between $50,000 and $100,000. His other assets amounted to no more than $170,000. He remained at a similar level until he became Speaker of the House. But by the time he set down the Speaker's gavel, he was substantially better off than when he entered office, with a reported net worth of up to $11 million.[3]*

A report from the *Business Insider* elaborates on the source of Hastert's new found wealth:

> *In 2005, Hastert purchased (or had a hand in purchasing) 264 acres near the site of the proposed "Prairie Parkway," and the site of a planned real estate development. Months after the purchases in early 2005, he*

placed a $207 million earmark into the federal high-
way bill to fund the parkway. He sold 69 acres months
later for $4.9 million—and netted between $2 mil-
lion and $10 million in a year.[4]

Whereas Hastert was willing to sell out his ideals for profit, tea party conservatives are now being accused of doing just the opposite. The phrase "purity for profit" has come to be used to demonize conservative organizations that want to elect leaders who will actually keep their promises, and resist the Beltway allure in a way that Hastert could not.[5]

Hastert retired in 2007 after handing over control of the House to Nancy Pelosi. Grassroots outrage over spending earmarks by Republicans helped fuel the shift in power.

Hastert was a cheater. The charms of Beltway power were just too compelling. Pelosi similarly used her position of power to the benefit of her portfolio of properties and investments. But her insider trading and self-dealing seems a better fit within the new Democratic Party, a party that has so fully embraced an expansive government on all aspects of our lives: in our health care, in choosing winners and losers on Wall Street, in expanding the power of the NSA and the IRS, and all the alphabet-soup agencies encroaching on your civil liberties, and even in expanding the war powers of the chief executive.

In 2008, this bipartisan collusion of insiders and politicians-for-life and special interests were busy driving America off a fiscal cliff. First it was Republicans, and then Democrats, but the real story was a mutual admiration club that saw politics as the end in itself, and policy as a by-product. It was a well-paid game, but the policy outcomes served only their interests, not ours.

So the American people rose up in protest, armed with new tools like Facebook, Twitter, Ning (the rotary phone of social networking), and RSS feeds that provided real-time information from bloggers and a multitude of disintermediated media sources. Previously disenfranchised voters were newly committing to get involved, to enter into a long-term relationship and a binding fidelity to first principles. And they were armed with freedom: a host of new online tools that lowered the barriers of entry to people trying to participate in the People's Business.

So began a permanent paradigm shift in American politics, shifting power from them to us.

And not a moment too soon. The bipartisan purveyors of business as usual seemed utterly uninterested in the consequences of their actions.

PARTY POLITICS

There was a time when I had higher aspirations for Republicans. I worked as the chief economist for Lee Atwater at the Republican National Committee. I was a foot soldier in the "Republican Revolution" of 1994, working for a Republican congressman as we sought to rein in a federal budget that was bleeding red ink.

I noticed over the years that the only great political successes enjoyed by Republicans were inexorably linked to a party that stood for something, that stood on principle. That's what had happened in 1994, when Republicans took control of the House of Representatives for the first time in forty years, based on a contractual promise to balance the budget and fix broken entitlements like welfare. And, yes, to stop a government takeover of our health care.

Of course the GOP takeover in 1995 would eventually devolve into business as usual, particularly under the Bush administration. Republicans passed the Patriot Act, an expansion of powers under the Foreign Intelligence Surveillance Act, an unsustainable increase in spending on a bankrupt Medicare program, and the practice of earmarking federal spending favors to preferred members of Congress, a practice that made a former high school wrestling coach turned Speaker a very wealthy man.

Spending and the size of government exploded under Republicans' watch, propelling the election of a little-known state legislator, Barack Obama, to the U.S. Senate in 2004, and the Pelosi Democrats to control of the House in 2006. Both Obama and Pelosi ran for office, against Republicans, promising renewed fiscal responsibility.

In 2006, Obama as the new senator from Illinois voted against increasing the debt ceiling, arguing:

> *Increasing America's debt weakens us domestically and internationally. Leadership means that "the buck stops here." Instead, Washington is shifting the burden of bad choices today onto the backs of our children and grandchildren. America has a debt problem and a failure of leadership. Americans deserve better.*[6]

He had a point. All of the spending and all of the borrowing was mortgaging the futures of future generations. It was generational theft. Of course there are limits. Even Washington can only spend so much money it does not have. You can only tax so much before producers revolt and stop generating new investment and new income to be taxed away at the margin. There is only so much that our government can borrow from the Chinese government. So what happened? Lots of easy money and credit issued by the Fed monetized all of the easy money being spent by Congress. That easy money fueled bad behavior on Wall Street, and the mega-banks bet it all knowing that someone would bail them out.

With the artificial boom came the inevitable bust in 2008.

The Democrats doubled down, providing the votes for the $700 billion TARP bailout (thank you, Senators Barack Obama [D-IL] and John McCain [D-DC]). President Obama, ignoring the promises he made during his courtship with your vote, proceeded to spend another $700 billion on crony-allocated "stimulus" on the failed projects of the politically connected. No one was afforded a chance to read what Washington was passing into law. So much for his concerns about "shifting the burden of bad choices today onto the backs of our children and grandchildren."

HEART AND SOUL

The first time I saw Ronald Reagan speak was at the White House in 1986. He quoted Ludwig von Mises. I was still a graduate student at the time and knew very little about Washington politics, but I thought it was pretty cool to hear the president of the United States quote my favorite economist. I thought, naïvely, that it was normal, representative of the Republican philosophy based on free enterprise, individual liberty, and a nation of boundless opportunity for those willing to work for it.

I wanted to stand with those guys.

I later learned that Reagan was never "normal," according to the political establishment. In 1965, the GOP establishment viewed Reagan, by then a candidate for governor in California, as a real threat. "G.O.P. Moderates Fear Coup by Reagan on Coast," read one *New York Times* headline. The former actor was "closely identified with the right-wing of the Republican Party."

Reagan's response? "I think basically that I stand for what the bulk of Americans stand for—dignity, freedom of the individual, the right to determine your own destiny."[7]

In 1975, Manny Klausner of *Reason* magazine asked Reagan about his political philosophy. The now former governor of California was equally succinct:

> *If you analyze it I believe the very heart and soul of conservatism is libertarianism. I think conservatism is really a misnomer just as liberalism is a misnomer for the liberals—if we were back in the days of the*

*Revolution, so-called conservatives today would be the
Liberals and the liberals would be the Tories. The basis
of conservatism is a desire for less government interfer-
ence or less centralized authority or more individual
freedom and this is a pretty general description also of
what libertarianism is.*[8]

In 1976, Reagan took on the GOP establishment again
by challenging, and almost toppling President Gerald Ford
at the Republican National Convention. Can you just
imagine John McCain's indignation, had he been a senior
Republican senator serving at the time? *"Who is this wacko
bird?"*

Like Ronald Reagan in 1976, today we may have to beat
the Republicans before we can beat the Democrats.

A LONG-TERM RELATIONSHIP

As I moved away from an academic career into public policy
and politics in the late 1980s, I started to discover just how
unusual it was to hear a politician speak, with credibility,
about the simple values of freedom that I had long ago discovered inside the cover fold of the Rush album called *2112*.
In Washington, D.C., there are too few people willing to
fight for the dignity of individual Americans, to stand un-
waveringly by commonsense rules that say *Don't Hurt People and Don't Take Their Stuff.*

So a little disruption of the status quo in Washington
seems like the only reasonable thing to do. It seems crazy to

do the same thing over and over and expect better results. It seems irrational to believe that the same closed, top–down leadership that got America into this mess offers any guidance on how to get America back on track toward more freedom and upward mobility and economic prosperity. It seems hopelessly naïve to think that the bipartisan collusion that continues to drive the growth of the national debt can now produce real solutions.

So, if they won't stand up, we should. Maybe it's time to make a long-term commitment to each other, to take on the serial cheaters in Washington, D.C. A devoted relationship would be so much better. A lifetime dedication to a set of principles that doesn't change based on the latest public opinion poll is just more satisfying, more fulfilling, more meaningful.

So, it's Them versus Us. There is no putting the genie back in the bottle; there's no stopping the newly empowered grass roots from reclaiming their property rights in the American enterprise. When the dust settles, historians will write about this political realignment as a tipping point in America, a paradigm shift that changed the rules. They will write about the ways that America finally beat Washington.

In markets, we now define our own experience online, relentlessly, even as savvy marketers attempt to influence our preferences. But regardless of how much they may know about our interests and habits, we still choose where to go, what to buy, and whom to ignore. Apply that same bottom-up independence to Washington insiders and the newly democratized political process and you can understand exactly why they are so totally, completely freaked out. They are freaking out because you know their number and the fixed rules of the game they are playing.

Republicans still uncomfortable outside the protective shell of their navy Brooks Brothers blazers are struggling to catch the technology wave that is so ubiquitously returning power, and knowledge, to individual shareholders. I've struggled to explain this difference, because so many things sound the same in the wake of the 2012 elections. With so much talk about the need to "engage the grass roots," "bridge the technology gap," and build the ultimate "Big Data" set coming from each corner of the right-of-center coalition, it's all getting jumbled together. From über-consultants like Karl Rove, to the dustiest of paper-churning think tanks, everyone is spouting the same talking points. The Republican National Committee even hired a chief technology officer (though only *after* the electoral drubbing of Mitt Romney.)

All of this is a blessing—an apparent rethinking of things that marks a critical reassessment of strategy and tactics among Republicans, conservatives, and libertarians. But these big "rethinks" are often one-dimensional. They get stuck on PR rehabilitation instead of serious self-reflection and a retooling of fundamentals. Remember the folks who tried to one-up Windows as a PC operating system when they could have been inventing the smartphone? Me neither. It goes to show, you can't just do what someone else did because they did it and it worked for them.

Shouldn't our strategy do more than ape the Big Data strategy of radical progressives? Can we learn from the left's effective use of mass personalization from the top down, and apply their technological savvy to a world that is becoming more decentralized, more democratized, and more free-to-choose?

The difference between them and us is simple: We are always in it for the long haul. Our proposition, contra the typical political pickup lines, is for a long-term relationship—a true fidelity to certain values and an unwavering commitment to each other. Why? Because one-night stands never work out, and our individual liberties cannot be defended by a single act, or better-behaved politicians.

We know that politicians respond to voter demand. Tangible expressions of consumer sentiment can also change the behavior of government bureaucrats, Republican precinct captains, members of school boards, and even Fortune 500 corporate CEOs. So, it is not enough to be steadfast to ideas. It is clearly never enough to show up once every couple of years on the first Tuesday in November. The days after Election Day matter more, because success at the ballot box can't translate into good public policy without the consistent demands of a constituency for economic freedom. This is Public Choice 101. One-night stands may achieve electoral success, but politicians will cheat on you when left to their own devices once they get to Washington, D.C. There are just too many temptations there, too many offers they can't refuse.

My New Tattoo

I have a new tattoo. The idea came to me as I worked on this book, and the symbolism seemed important to me. You see, tattoos last forever. Tattoos are permanent, and if you are going to get one, you need to know what you're doing. You have to be committed to it. You had better be in it for the long haul. Otherwise, don't do it.

I have a friend, Joel, who has become an integral part of the same growing community that I belong to, part of a new generation of citizen freedom fighters. Joel calls us the "Liberty Movement." We have worked together on the ground in Ohio, organizing, fighting, sometimes winning. Joel is a small business guy—an entrepreneur—who owns and operates Marv's Place with his wife, Danielle. He's way too busy with his first responsibilities to his job and his family to be dedicating so much time to citizen activism. But he makes time.

Although Joel was born with congenital scoliosis, a sideways curvature of the spine, and has endured years of surgeries and invasive medical procedures, he has never let it slow him down. In 2013, Joel finished in an astonishingly close second place for his local city council election, losing by just eleven votes to a former mayor with far more resources and far better political connections. He's already planning for his next run.

"Danielle and I live on a pretty limited income and have to watch what we spend," Joel once told me. "As most couples do, we were discussing our bills, finances, and discretionary spending together and how I was spending quite a bit of time and money traveling to meetings and events. Danielle was worried about it, and asked me if we could afford all that I put into the fight for freedom."

"How can we can we afford not to?" Joel asked back.

JOEL DECIDED TO GET the FreedomWorks star tattooed on his forearm. It's kinda badass. And it tells me that Joel Davis is in it for the long haul.

So, as I was writing my new book, I worked on new ink. Stealing inspiration from my friend Joel, the right tattoo seemed like a perfect metaphor for our fight to be free. It's not a one-time thing. It will be there tomorrow, and next week, until the day I die.

Early on in research for this book, I found a great little essay by "An American Guesser," published in 1775. The real author is none other than Benjamin Franklin. If Thomas Jefferson was the idealist, and George Washington the leader, and James Madison the architect, Samuel Adams the community organizer, then Franklin would have been the Yoda of the founding generation. But he was also a journalist who had a way of translating deeply held values into a good story. This particular one is an allegory about the rattlesnake.

I recollected that her eye excelled in brightness, that of any other animal, and that she has no eye-lids. She may therefore be esteemed an emblem of vigilance. She never begins an attack, nor, when once engaged, ever surrenders: She is therefore an emblem of magnanimity

and true courage. As if anxious to prevent all preten-
sions of quarreling with her, the weapons with which
nature has furnished her, she conceals in the roof of her
mouth, so that, to those who are unacquainted with
her, she appears to be a most defenseless animal; and
even when those weapons are shown and extended for
her defense, they appear weak and contemptible; but
their wounds however small, are decisive and fatal.
Conscious of this, she never wounds 'till she has gen-
erously given notice, even to her enemy, and cautioned
him against the danger of treading on her.[9]

Franklin believed that the rattlesnake reflected "a strong picture of the temper and conduct of America." Mind your own business. Don't hurt others. If attacked, never back down. Fight the power when tread upon.

'Tis curious and amazing to observe how distinct and
independent of each other the rattles of this animal are,
and yet how firmly they are united together, so as never
to be separated but by breaking them to pieces. One of
those rattles singly, is incapable of producing sound, but
the ringing of thirteen together, is sufficient to alarm
the boldest man living.

Franklin, speaking to Americans struggling to come together in common purpose against a grave external threat, anticipates the profound strength of closely knit communities that respect the individual rights of their constituent members. Together, in voluntary association, we can accomplish great things.

> *The Rattle-Snake is solitary, and associates with her kind only when it is necessary for their preservation. In winter, the warmth of a number together will preserve their lives, while singly, they would probably perish. The power of fascination attributed to her, by a generous construction, may be understood to mean, that those who consider the liberty and blessings which America affords, and once come over to her, never afterwards leave her, but spend their lives with her.*

As you may have guessed, my new tattoo is a rattlesnake. It says "Join, Or Die."

Like Joel, and many millions of other newly engaged Americans, I'm all in. Ben Franklin and his partners in liberty were all in too. They bet it all on an idea. An idea few "experts" believed would work. They signed their "John Hancock" on that parchment in defense of the idea that individuals are free, that free individuals do not serve government ends, and that governments exist only to the extent that we the stakeholders permit it so.

"*That to secure these rights, Governments are instituted among Men, deriving their just powers from the consent of the governed . . .*"

George Washington, in his inaugural proposal to each of us, offered that "[t]he preservation of the sacred fire of liberty, and the destiny of the republican model of government, are justly considered as deeply, perhaps as finally staked, on the experiment entrusted to the hands of the American people."

He was seeking your hand in a long-term relationship, right? He was asking us all to choose the burden of commitment, with full knowledge that it wouldn't always be easy, that the weight of responsibility for a successful relationship falls on your shoulders first.

The fight for liberty is a burden that requires eternal vigilance. You have to work at it. You will be there, for liberty, in good times and in bad. Even if the IRS targets your efforts to gather your neighbors in peaceful protest, or punishes you for petitioning your government representatives for a redress of grievances. Some extraordinary soul might have to commit to keep speaking out for equal treatment under the law even when faceless, gray-suited bureaucrats imbedded deep within the FBI deem you "the most dangerous negro in America." You might be targeted, simply because you stood up and spoke out, calling for all Americans to be judged based on the content of their character. You too may have to pledge everything, including your life, your fortune, and your honor.

Our fight, unlike politics, is all about the long term. Can we continue to build community that will be there for the long haul? Can we come together as a beautiful mess of individual aspirations bonded by a shared set of values.

New technologies and the decentralization of news and information are shifting power away from Washington insiders to citizens, and this paradigm shift is in direct conflict with Beltway efforts to reconsolidate power and control information and behavior, from the top down. But try as they might, I don't think they can stop us from reclaiming what is rightfully ours.

Isn't this exactly the American way? Bottom-up governance based on the rule of law, originating from engaged, ever-vigilant citizens, channeled through an accountable legislature, to the chief executive's desk. We are the shareholders. We don't believe in czars, governance by midnight order, or the expansive power of the executive branch. It doesn't matter who the president may be. He or she will always report to us.

The weight of liberty is a burden. It's a lifetime commitment. But the upside is so awesome.

ACKNOWLEDGMENTS

ON SEPTEMBER 4, 2012, I was met in my office by an armed guard. "Who are you?" I asked. "Do you work for Freedom-Works?" He refused to identify himself, and I refused to hand over my iPhone. With that, Executive Vice President Adam Brandon and I were perp-walked out of Freedom-Works headquarters.

Thus began the seventy-two-hour occupation of FreedomWorks, a surreal hostile takeover bid by three Board members with close ties to the GOP establishment. I did not see this coming—I should have—and we all paid a price for that.

As political intrigue, this bizarre, *House of Cards*–like episode was probably quite typical: it was all about personal betrayal, money, and power. What was anything but typical, particularly in Washington, D.C., was the ironclad unity of the FreedomWorks "family," without which this book, or the continued existence of our organization, would not have been possible.

It might have been easier to back down, to walk away, but I was bucked up by the unbending commitment of my colleagues and eight steadfast members of our Board of Directors. The "family" stood together, even as many of them were fired, sometimes multiple times. What an honor it is to work with such people.

Others that did not need to stand with us did so without hesitation. My friend Glenn Beck was one, but there were many others who took a stand, and a few bullets, in this fight for the heart and soul of FreedomWorks.

Equally important, of course, is the resolute dedication of the grassroots community we serve. They set the bar and represent an existential threat to the D.C. power structure, and the establishment knows it. The community's resolve in the face of all the adversity, the long odds, and way too much "friendly" fire, is inspirational. I think we are all figuring out the rules of the game together. Knowing is liberating, even when knowledge comes at a premium.

The attempted coup was the fire that ignited the writing of this book, based on the theory that things that don't kill you can make you stronger. I think that's true. As I have before, I fed off of the insights of the late, great moral philosopher Warren Zevon. "I'll sleep when I'm dead," he once sang.

Mostly, and for everything that is important in my life, my Unmoved Motivator is my awesome wife and partner in life, Terry. This book, and most things, would not have happened without her. She keeps me straight, focused, and usually headed in the right direction. Despite this daunting challenge, Terry still hasn't given up on me after twenty-seven years. It may be too late to get out now, Mama Bear.

For this project there were a number of people essential to its success. Peter Hubbard at HarperCollins was an unwavering and reliable hand guiding *Don't Hurt People* from concept to publication. He was a champion of the project from day one. This is the third book we have worked on together, and it's always a great experience.

Senators Mike Lee, Rand Paul and Ted Cruz all took time out of their insanely busy schedules to talk to me. So did Representatives Justin Amash, Thomas Massie, and David Schweikert. It is great, and somewhat disorienting, to have so many principled politicians that I might have included in this book, but these six were the obvious best choices. None of them, of course, are responsible for the crazy rantings in this book, except where they are directly quoted.

Joel Davis, unrelenting fighter for liberty, took time to retell his story for me. We are currently conspiring over the next tattoo.

Adam Brandon was a key player as well, at least when he wasn't getting fired. He would agree to unreasonable deadlines and then hand it off to me. He does that best. We are working on his grasp of the economic concept of opportunity cost: *If you do this, you can't do that.*

Logan Albright contributed substantial research during the writing of this book, and served as traffic cop for all of the various inputs. Wayne Brough and Reid Smith also contributed. Laura Howd ensured that the trains ran on time and deadlines were met. Logan organized all citations and made sure the footnotes were in proper form, which was a daunting task. Laura, Josh Withrow, Dean Clancy, Emilia Huneke-Bergquist, Jackie Bodnar, Easton Randall, Andrew

Smith, Parissa Sedghi, Kara Pally, and Christine Domenech all read the final manuscript for mistakes, catching many. Any remaining ones are mine, of course.

As he did during the writing marathons for *Give Us Liberty* and *Hostile Takeover*, Roark the cat played a key role as a calm presence, chooser of music, and random, but critical, keyboard adjustments. Some of his preferred settings on my iMac appear to be permanent. It was his idea to include so much Ayn Rand in *Don't Hurt People*, although the one Howard Roark quote

My co-author, Roark the Cat.

somehow ended up on the cutting-room floor. He must have been napping.

Speaking of music, a lot of Rush was played during the writing of the book, as you may have ascertained by the end of Chapter Two. If you don't already have it, you should get a vinyl copy of *A Farewell to Kings*. Liner notes are everything. The band members will no doubt be horrified to learn of their unwilling role in this process, and I can only assume that *Rolling Stone* will demand a(nother) clarification from the band. There was also plenty of John Coltrane, Tom Waits, My Morning Jacket, Sigur Ros, and Father John Misty. And, of course, the Grateful Dead.

Finally, a random hat tip to Three Floyds Brewing's *Permanent Funeral*. It's a fine beer that fortified this author at key junctures in the creative writing process.

NOTES

Chapter 1: Rules for Liberty

1. Sir John Emerich Edward Dalberg-Acton, Lecture, February 26, 1877.
2. Murray Rothbard, "War, Peace, and the State," *The Standard* (April 1963), 2–5.
3. Adam Smith, *The Theory of Moral Sentiments* (New York: Penguin Classics, 2010).
4. Ibid.
5. Max Weber, *Economy and Society* (Berkeley: University of California Press, 1978).
6. Wendy Milling, "President Obama Jabs at Ayn Rand, Knocks Himself Out," *Forbes,* October 30, 2012, http://www.forbes.com/sites/real spin/2012/10/30/president-obama-jabs-at-ayn-rand-knocks-himself-out/ (accessed October 23, 2013).
7. Thomas Patrick Burke, "The Origins of Social Justice: Taparelli d'Azeglio," *First Principles Journal,* January 1, 2008, http://www.firstprinciples journal.com/articles.aspx?article=1760 (accessed August 21, 2013).
8. John Rawls, *A Theory of Justice* (Oxford: Oxford University Press, 1971).
9. F. A. Hayek, *The Fatal Conceit: Errors of Socialism* (London: Routledge, 1988), 114.
10. Jim Geraghty, "The Things We Choose to Do Together," *National Review,* August 27, 2008, http://www.nationalreview.com/campaign -spot/8984/things-we-choose-do-together (accessed September 17, 2013).

11. Stephen Cruz, "Ashton Kutcher Reveals to Millennials an Enduring Secret to Certain Success," *Forbes,* August 23, 2013, http://www.forbes.com/sites/realspin/2013/08/23/ashton-kutcher-reveals-to-milennials-an-enduring-secret-to-certain-success/ (accessed October 23, 2013).

12. Ludwig von Mises, *Human Action* (New Haven, CT: Yale University Press, 1949), 859.

13. Jonathan Haidt, "What the Tea Partiers Really Want," *Wall Street Journal,* October 16, 2010, http://online.wsj.com/news/articles/SB10001424052748703673604575550243700895762 (accessed December 15, 2013).

14. Benjamin A. Rogge, *Can Capitalism Survive?* (Indianapolis: Liberty Fun, Inc., 1979).

15. Sir John Emerich Edward Dalberg-Acton, Letter to Mendell Creighton, April 5, 1887.

16. F. A. Hayek, *The Constitution of Liberty* (Chicago: University of Chicago Press, 1960).

17. Ibid.

Chapter 2: You Can't Have Freedom for Free

1. Ron Wynn, Michael Erlewine, and Vladimir Bogdanov, *The All Music Guide to Jazz* (San Francisco: Miller Freeman Books, 1994), 197.

2. *Rush: Classic Albums: 2112 & Moving Pictures* (Eagle Rock Entertainment, 2010).

3. Chris Matthew Sciabarra and Larry J. Sechrest, "Ayn Rand Among the Austrians," *Journal of Ayn Rand Studies* 6, no. 2 (Spring 2005): 241–50.

4. Pete is now one of the most distinguished professors in a booming community of Austrian-minded economists teaching in academia, and is the deputy director of the James M. Buchanan Center for Political Economy, a Senior Research Fellow at the Mercatus Center, and a professor in the economics department at George Mason University.

5. H. L. Mencken Quotes, Goodreads, http://www.goodreads.com/author/quotes/7805.H_L_Mencken?page=3 (accessed August 29, 2013).

6. Barry Miles, "Is Everybody Feelin' All RIGHT? (Geddit?)," *New Musical Express,* March 4, 1978, http://cygnus-x1.net/links/rush/images/books/mojo-06.2012/mojo-06.2012-11.pdf (accessed September 26, 2013).

7. Scott R. Benarde, "How the Holocaust Rocked Rush Front Man Geddy Lee," *JWeekly,* June 25, 2004, http://www.jweekly.com/article/full/23003/how-the-holocaust-rocked-rush-front-man-geddy-lee/ (accessed August 29, 2013).

8.   Stephen Cox, "Ayn Rand's Anthem: An Appreciation," Atlas Society, http://www.atlassociety.org/ayn-rands-anthem-appreciation (accessed August 29, 2013).

9.   "Books That Made a Difference in Readers' Lives," English Companion, response to Survey of Lifetime Reading Habits taken in 1991, http://www.englishcompanion.com/Readings/booklists/loclist.html (accessed August 29, 2013).

10.   Joshua Green, "An Apology to Rand Paul," *The Atlantic,* June 11, 2010, http://www.theatlantic.com/entertainment/archive/2010/06/an-apology-to-rand-paul/57999/ (accessed August 29, 2013).

11.   Andy Greene, "Q&A: Neil Peart on Rush's New LP and Being a 'Bleeding Heart Libertarian,' " *Rolling Stone,* June 12, 2012, http://www.rollingstone.com/music/news/q-a-neil-peart-on-rushs-new-lp-and-being-a-bleeding-heart-libertarian-20120612#ixzz2dIlNJFqD (accessed August 29, 2013).

Chapter 3: Them Versus Us

1.   Martin Luther King, "I Have A Dream," speech delivered August 28, 1963, at the Lincoln Memorial in Washington, D.C., http://www.americanrhetoric.com/speeches/mlkihaveadream.htm (accessed September 3, 2013).

2.   William Sullivan, "Communist Party, USA Negro Question," U.S. Government Memorandum, August 30, 1963, http://americanradioworks.publicradio.org/features/king/images/fbifiles/other/full/10.jpg (accessed September 18, 2013).

3.   Tony Capaccio, "King Address That Stirred World Led to FBI Surveillance," *BloombergBusinessweek,* August 27, 2013, http://www.businessweek.com/news/2013-08-27/king-address-that-stirred-world-led-to-fbi-surveillance-program (accessed September 3, 2013).

4.   David J. Garrow, "The FBI and Martin Luther King," *The Atlantic,* July 1, 2002, http://www.theatlantic.com/magazine/archive/2002/07/the-fbi-and-martin-luther-king/302537/ (accessed September 3, 2013).

5.   Peter Hamby, "Axelrod Suggests Tea Party Movement Is 'Unhealthy,' " CNN, April 19, 2009, http://politicalticker.blogs.cnn.com/2009/04/19/axelrod-suggests-tea-party-movement-is-unhealthy/ (accessed September 16, 2013).

6.   Abby D. Phillip, "IRS Planted Question About Tax Exempt Groups," ABC News, May 17, 2013, http://abcnews.go.com/blogs/politics/2013/05/irs-planted-question-about-tax-exempt-groups/ (accessed September 3, 2013).

7.   Victor Fleischer, "A Dickensian Delay at the IRS," *New York Times,*

May 16, 2013, http://dealbook.nytimes.com/2013/05/16/a-dickensian -delay-at-the-i-r-s/ (accessed September 19, 2013).

8. David Weigel, "IRS Asked a Pro-Life Group to Explain Its Prayers Outside Planned Parenthood, Which Is Now a Scandal," *Slate,* May 17, 2013, http://www.slate.com/blogs/weigel/2013/05/17/the_irs_asked _a_pro_life_group_to_explain_its_prayers_outside_planned_parent hood.html (accessed September 26, 2013).

9. Tamara Keith, "Report: IRS Scrutiny Worse for Conservatives," NPR, July 30, 2013, http://www.npr.org/blogs/itsallpolitics/2013/07/30/ 207080580/report-irs-scrutiny-worse-for-conservatives (accessed August 26, 2013).

10. "IRS Scrutiny of Non-profit Organizations," C-SPAN video, June 4, 2013, http://www.c-spanvideo.org/program/RSScr (accessed August 29, 2013).

11. Ibid.

12. Caroline May, "Tea Party Groups Speak Out Against the IRS: 'Folks, This Is Bad,'" *Daily Caller,* May 16, 2013, http://dailycaller .com/2013/05/16/tea-party-groups-speak-out-against-the-irs-folks -this-is-bad/ (accessed August 29, 2013).

13. Michael McDonald, "The IRS Takes a Closer Look at Colleges," *BloombergBusinessweek,* November 17, 2011, http://www.business week.com/magazine/the-irs-takes-a-closer-look-at-colleges-11172011 .html (accessed September 16, 2013).

14. Stan Veuger, "Yes, IRS Harassment Blunted the Tea Party Ground Game," American Enterprise Institute, June 20, 2013, http://www .aei.org/article/economics/yes-irs-harassment-blunted-the-tea-party -ground-game/ (accessed September 18, 2013).

15. Douglas M. Charles, "How Did the IRS Get Investigatory Authority Anyway?," History News Network, August 21, 2013, http://hnn.us/article /151970#sthash.9RXzbkmd.dpuf (accessed September 16, 2013).

16. Michael Scherer, "New IRS Scandal Echoes a Long History of Political Harassment," *Time,* Swampland, May 14, 2013, http://swampland .time.com/2013/05/14/anger-over-irs-audits-of-conservatives -anchored-in-long-history-of-abuse/#ixzz2eR8AInHE (accessed September 16, 2013).

17. Alan Farnham, "IRS Has Long History of Political Dirty Tricks," ABC News, May 15, 2013, http://abcnews.go.com/Business/irs-irs-long -history-dirty-tricks/story?id=19177178 (accessed September 3, 2013).

18. Victor Reuther and Walter Reuther, "The Reuther Memorandum: The Radical Right in America Today," Memorandum to the Attorney General of the United States, December 19, 1961, http://www.scribd .com/doc/31124491/The-Reuther-Memorandum-Precusor-to-the -Ideological-Organizations-Audit-Project-Created-by-President-John-F

-Kennedy-and-Attorney-General-Robert-Kenn (accessed September 18, 2013).

19. David Dykes, "Former IRS Chief Recalls Defying Nixon," *USA Today,* May 26, 2013, http://www.usatoday.com/story/news/nation/2013/05/26/irs-chief-defied-nixon/2360951/ (accessed September 3, 2013).

20. House Judiciary Committee, "Articles for Impeachment," Watergate.info, July 27, 1974, http://watergate.info/impeachment/articles-of-impeachment (accessed September 16, 2013).

21. Scherer, "New IRS Scandal Echoes a Long History of Political Harassment."

22. "Lois Lerner to 1996 US Senate Candidate Al Salvi: 'We'll Get You!' " *Illinois Review,* June 3, 2013, http://illinoisreview.typepad.com/illinoisreview/2013/06/lerner-asked-salvi-for-200000-plus-never-run-again-promise.html (accessed September 16, 2013).

23. Josh Hicks, "Republicans Say IRS E-mails from Lois Lerner Show 'Abuse of Power,' " *Washington Post,* September 13, 2013, http://www.washingtonpost.com/blogs/federal-eye/wp/2013/09/13/republicans-say-irs-e-mails-from-lois-lerner-show-abuse-of-power (accessed September 16, 2013).

24. Ibid.

25. Timothy P. Carney, "The IRS Is Deeply Political and Very Democratic," *Washington Examiner,* May 15, 2013, http://washingtonexaminer.com/tim-carney-the-irs-is-deeply-political-and-very-democratic/article/2529758 (accessed September 16, 2013).

26. "IRS Chief Says He'd Rather Not Switch to ObamaCare Plan," Fox News, August 1, 2013, http://www.foxnews.com/politics/2013/08/01/irs-chief-says-hed-rather-not-switch-over-to-obamacare-plan (accessed September 3, 2013).

27. National Treasury Employees Union action alert, http://capwiz.com/nteu/issues/alert/?alertid=62634726&type=CO&utm_source=Illinois+Policy+Institute&utm_campaign=7790111647-0613_ecompass&utm_medium=email&utm_term=0_0f5a22f52c-7790111647-10830129 (accessed September 19, 2013).

28. Wenton Hall, "IRS Org Chart Puts Ingram, Lerner at Center of Power," Breitbart.com, May 23, 2013, http://www.breitbart.com/Big-Government/2013/05/23/EXCLUSIVE-IRS-Org-Chart-Puts-Ingram-and-Lerner-At-Center-of-Power (accessed September 16, 2013).

29. "Members Only: How the White House Is Weaseling Congress Out of ObamaCare," *Wall Street Journal,* August 7, 2013, http://online.wsj.com/article/SB10001424127887324522504578654193173779414.html (accessed August 27, 2013).

30. Historical Federal Workforce Tables, U.S. Office of Personnel Management: Data, Analysis & Documentation, http://www.opm.gov/policy

-data-oversight/data-analysis-documentation/federal-employment
-reports/historical-tables/total-government-employment-since-1962
(accessed August 26, 2013).

31. Ruth Alexander, "Which Is the World's Biggest Employer?," BBC News,
March 19, 2012, http://www.bbc.co.uk/news/magazine-17429786 (accessed August 26, 2013).

32. "IRS Scrutiny of Non-profit Organizations," C-SPAN, June 4, 2013.

Chapter 4: Gray-Suited Soviets

1. Jehiel Keeler Hoyt, *The Cyclopedia of Practical Quotations* (1896),
763.

2. "Obama on NSA Surveillance: Can't Have 100% Security and 100%
Privacy," RT.com, June 7, 2013, http://rt.com/usa/obama-surveillance
-nsa-monitoring-385 (accessed September 9, 2013).

3. Tal Kopan, "Lindsey Graham 'Glad' NSA Tracking Phones," Politico,
June 6, 2013, http://www.politico.com/story/2013/06/lindsey-graham
-nsa-tracking-phones-92330.html (accessed September 9, 2013).

4. Ludwig von Mises, *Human Action* (New Haven, CT: Yale University
Press, 1949), 283.

5. Sarah Kliff, "White House Delays Employer Mandate Requirement
until 2015," *Washington Post,* July 2, 2013, http://www.washingtonpost
.com/blogs/wonkblog/wp/2013/07/02/white-house-delays-employer
-mandate-requirement-until-2015 (accessed September 26, 2013).

6. Megan R. Wilson, "ObamaCare's Architects Reap Windfall as Washington Lobbyists," *The Hill,* August 25, 2013, http://thehill.com/
business-a-lobbying/318577-architects-of-obamacare-reap-windfall
-as-washington-lobbyists (accessed September 23, 2013).

7. Elijah E. Cummings and Sander M. Levin, "Reform the IRS, but
Leave Politics out of It," *Washington Post,* August 12, 2013, http://www
.washingtonpost.com/opinions/reform-the-irs-but-leave-politics-out
-of-it/2013/08/12/64c5d36c-0362-11e3-9259-e2aafe5a5f84_story
.html (accessed September 23, 2013).

8. Political Calculations, "How Many Pages Long Is the U.S. Income Tax
Code in 2013?," *Town Hall,* Finance, February 17, 2013, http://finance
.townhall.com/columnists/politicalcalculations/2013/02/17/how
-many-pages-long-is-the-us-income-tax-code-in-2013-n1514277 (accessed August 23, 2013).

9. Kelly Phillips Erb, "Tax Code Hits Nearly 4 Million Words, Taxpayer
Advocate Calls It Too Complicated," *Forbes,* January 10, 2013, http://
www.forbes.com/sites/kellyphillipserb/2013/01/10/tax-code-hits

-nearly-4-million-words-taxpayer-advocate-calls-it-too-complicated/ (accessed August 23, 2013).

10. Guinness World Records, http://www.guinnessworldrecords.com/ records-1/longest-novel/ (accessed August 23, 2013).

11. Patricia Murphy, "IRS Commissioner Does Not Do His Own Taxes," *Politics Daily,* January 8, 2010, http://www.politicsdaily.com/2010/ 01/12/irs-commissioner-admits-he-does-not-do-his-own-taxes (accessed August 23, 2013).

12. Janet Novack, "Tax Waste: 6.1 Billion Hours Spent Complying with Federal Tax Code," *Forbes,* January 5, 2011, http://www.forbes .com/sites/janetnovack/2011/01/05/tax-waste-6-1-billion-hours-spent -complying-with-federal-tax-code (accessed September 9, 2013).

13. Jason J. Fichtner and Jacob Feldman, "The Hidden Costs of Tax Compliance," Mercatus Center, May 20, 2013, http://mercatus.org/publica tion/hidden-costs-tax-compliance (accessed September 9, 2013).

14. Leslie Bonacum and Eric Scott, "When It Comes To Tax Law, It's Complicated," CCH, January 2012, http://www.cch.com/wbot2012/ 020TaxCode.asp (accessed September 9, 2013).

15. Alexandra Wexler, "Sugar Free Trade Sours for Taxpayers," *Wall Street Journal,* November 19, 2013, http://blogs.wsj.com/moneybeat/ 2013/11/19/sugar-free-trade-sours-for-taxpayers (accessed December 15, 2013).

16. Wendy McElroy, "Decriminalize the Average Man," Ludwig von Mises Institute, October 12, 2010, http://mises.org/daily/5759 (accessed August 23, 2013).

17. Avik Roy, "How Employer-Sponsored Insurance Drives Up Health Costs," *Forbes,* May 12, 2012, http://www.forbes.com/sites/aroy/2012/ 05/12/how-employer-sponsored-insurance-drives-up-health-costs (accessed September 9, 2013).

18. Center for Consumer Information and Insurance Oversight, "Annual Limits Policy: Protecting Consumers, Maintaining Options, and Building a Bridge to 2014," Centers for Medicare and Medicaid Services, January 6, 2012, http://www.cms.gov/CCIIO/Resources/ Files/approved_applications_for_waiver.html (accessed September 18, 2013).

19. Becket Adams, "How Many Are Exempt? The Final Number of 'ObamaCare' Waivers Is In," *The Blaze,* January 6, 2012, http://www.the blaze.com/stories/2012/01/06/how-many-businesses-are-exempt -the-final-number-of-obamacare-waivers-is-in (accessed September 3, 2013).

20. Alex Nussbaum, "ObamaCare Unleashes Benefit Changes from Companies," Bloomberg, September 19, 2013, http://www.bloomberg.com/

news/2013-09-19/obamacare-unleashes-benefit-changes-from-compa
nies.html (accessed September 23, 2013).

21. Avik Roy, "The Obamacare Exchange Scorecard: Around 100,00 Enrollees and Five Million Cancellations," *Forbes,* November 12, 2013,
http://www.forbes.com/sites/theapothecary/2013/11/12/the-obama
care-exchange-scorecard-around-100000-enrollees-and-five-million
-cancell ations (accessed December 5, 2013).

22. Mike Emmanuel, "Second Wave of Health Plan Cancellations Looms,"
Fox News, November 20, 2013, http://aei.org/article/health/second
-wave-of-health-plan-cancellations-looms (accessed December 5,
2013).

23. "Members Only: How the White House Is Weaseling Congress Out
of ObamaCare," *Wall Street Journal,* August 7, 2013, http://online
.wsj.com/article/SB100014241278873245225045786541931737794
14.html (accessed August 27, 2013).

24. Glenn Kessler, "Did Obama Exempt 1,200 Groups, Including Congress, from Obamacare?" *The Washington Post,* October 16, 2013,
http://www.washingtonpost.com/blogs/fact-checker/wp/2013/10/16/
did-obama-exempt-1200-groups-including-congress-from-obamacare
(accessed December 12, 2013).

25. Jennifer G. Hickey, "Democrats Act to Stop Vitter Amendment, Keep
ObamaCare Exemptions for Congress," *NewsMax,* September 17, 2013,
http://www.newsmax.com/Newsfront/vitter-obamacare-congress
-exemptions/2013/09/17/id/526269 (accessed September 23, 2013).

26. Avik Roy, "Yet Another White House ObamaCare Delay: Out-Of
-Pocket Caps Waived Until 2015," *Forbes,* August 13, 2013, http://
www.forbes.com/sites/theapothecary/2013/08/13/yet-another-white
-house-obamacare-delay-out-of-pocket-caps-waived-until-2015 (accessed September 9, 2013).

27. Internal Revenue Service Oversight Hearing, C-SPAN, April 9, 2013,
http://www.c-spanvideo.org/program/OversightHearing (accessed September 19, 2013).

28. Robert E. Barnes on behalf of John Doe Company et al., Class Action Complaint, Superior Court of the State of California for San Di-
ego, March 11, 2013, http://global.nationalreview.com/pdf/complaint
_051513.pdf (accessed August 19, 2013).

29. Internal Revenue Service Oversight Hearing, C-SPAN.

30. Jake Sherman, "Nancy Pelosi Says She's 'Proud' of Obamacare," *Politico,* October 30, 2013, http://www.politico.com/story/2013/10/nancy
-pelosi-says-shes-proud-of-obamacare-99102.html (accessed December
5, 2013).

31. "Transcript of Obama's Announcement on Health Insurance," *Wall
Street Journal,* November 14, 2013, http://blogs.marketwatch.com/

capitolreport/2013/11/14/transcript-of-obamas-announcement-on
-health-insurance (accessed December 12, 2013).

32. Avik Roy, "HHS Inspector General: Obamacare Privacy Protections
Way Behind Schedule; Rampant Violations of Law Possible," *Forbes,*
August 7, 2013, http://www.forbes.com/sites/theapothecary/2013/
08/07/hhs-inspector-general-obamacare-privacy-protections-way
-behind-schedule-rampant-v iolations-of-law-possible (accessed September 9, 2013).

33. John Merline, "Think NSA Spying Is Bad? Here Comes ObamaCare
Hub," *Investor's Business Daily,* June 25, 2013, http://news.investors
.com/062513-661264-obamacare-database-hub-creates-privacy-night
mare.htm (accessed September 9, 2013).

34. *Identity Theft Resource Center,* "2013 Data Breach Stats," August 9,
2013, http://www.idtheftcenter.org/images/breach/Breach_Stats_
Report_2013.pdf (accessed December 5, 2013).

35. Treasury Inspector General for Tax Administration, "Some Taxpayers
Were Not Appropriately Notified When Their Personally Identifiable
Information Was Inadvertently Disclosed," May 24, 2011, http://www
.treasury.gov/tigta/auditreports/2011reports/201140054fr.pdf (accessed September 9, 2013).

36. Thomas Hargrove, "Social Security Kept Silent About Private Data
Breach," *Seattle Times,* October 13, 2011, http://seattletimes.com/
html/nationworld/2016498264_socialsecurity14.html (accessed September 9, 2013).

37. Merline, "Think NSA Spying Is Bad? Here Comes ObamaCare Hub."

38. Patrick Meehan and James Lankford, "A Closer Look at the ObamaCare
Data Hub," *The Hill,* August 2, 2013, http://thehill.com/blogs/con
gress-blog/healthcare/315083-a-closer-look-at-the-obamacare-data
-hub (accessed September 9, 2013).

39. "Your Next IRS Political Audit," *Wall Street Journal,* May 14, 2013,
http://online.wsj.com/article/SB100014241278873247157045784814
61934680982.html (accessed September 9, 2013).

40. Department of Health and Human Services, "Notice to Establish a New
System of Records," *Federal Register,* vol. 78, no. 25, February 6, 2013,
http://www.gpo.gov/fdsys/pkg/FR-2013-02-06/html/2013-02666
.htm (accessed September 9, 2013).

41. Ashton Ellis, "ObamaCare's 'Data Hub' Should Be Its Death Knell,"
Center for Individual Freedom, July 25, 2013, http://cfif.org/v/index
.php/commentary/56-health-care/1910-obamacares-data-hub-should
-be-its-death-knell (accessed September 9, 2013).

42. John Fund, "ObamaCare's Branch of the NSA," *National Review,*
July 22, 2013, http://www.nationalreview.com/article/354031/obama
cares-branch-nsa-john-fund (accessed September 9, 2013).

Chapter 5: Same as the Old Boss

1. James M. Buchanan and Richard E. Wagner, *Democracy in Deficit: The Political Legacy of Lord Keynes* (Indianapolis: Liberty Fund, Inc. 1999).

2. John Maynard Keynes, *A Tract on Monetary Reform* (London: MacMillan & Co. 1923).

3. Thomas Schatz, "Putting the National Debt in Perspective," *Daily Caller,* September 19, 2012, http://dailycaller.com/2012/09/19/putting-the-national-debt-in-perspective (accessed September 27, 2013).

4. Steve Hargreaves, "Labor Participation Lowest Since 1978," *CNN Money,* September 6, 2013, http://money.cnn.com/2013/09/06/news/economy/labor-force-participation/index.html (accessed September 27, 2013).

5. Bureau of Labor Statistics, "Table A-10. Selected Unemployment Indicators, Seasonally Adjusted," http://www.bls.gov/news.release/empsit.t10.htm (accessed September 27, 2013).

6. Richard Fry, "A Rising Share of Young Adults Live in Their Parents' Home," Pew Research, August 1, 2013, http://www.pewsocialtrends.org/2013/08/01/a-rising-share-of-young-adults-live-in-their-parents-home (accessed September 27, 2013).

7. Heidi Moore, "U.S. Student Loan Debt by the Numbers," *Guardian,* April 2, 2013, http://www.theguardian.com/money/2013/apr/03/student-loan-debt-america-by-the-numbers (accessed September 27, 2013).

8. Joe Light, "Many Graduates Delay Job Searches," *Wall Street Journal,* June 3, 2013, http://online.wsj.com/article/SB10001424052702303657404576363783070164132.html (accessed September 27, 2013).

9. Donghoon Lee, "Household Debt and Credit: Student Debt," Federal Reserve Bank of New York, February 28, 2013, http://www.newyorkfed.org/newsevents/mediaadvisory/2013/Lee022813.pdf (accessed August 26, 2013).

10. "Average Net Price for Full-Time Students over Time—Public Institutions," Trends in Higher Education, http://trends.collegeboard.org/college-pricing/figures-tables/average-net-price-full-time-students-over-time-public-institutions (accessed September 27, 2013).

11. Alec Liu, "The Student Loan Bubble Looks Awfully Like the Housing Crisis, Top Bankers Say," *Motherboard,* May 13, 2013, http://motherboard.vice.com/blog/the-student-loan-bubble-looks-awfully-like-the-housing-crisis-bankers-warn-fed (accessed August 26, 2013).

12. Roper Center, "How Groups Voted in 2008," http://www.ropercenter.uconn.edu/elections/how_groups_voted/voted_08.html; and "How

Groups Voted in 2012," http://www.ropercenter.uconn.edu/elections/how_groups_voted/voted_12.html (accessed September 26, 2013).

13. FreedomWorks, "The Role of Government," September 2013, http://online-campaigns.s3.amazonaws.com/docs/pollingreport.pdf (accessed September 12, 2013).

14. "Millennials: Confident. Connected. Open to Change," Pew Research Center, February 24, 2010, http://www.pewsocialtrends.org/2010/02/24/millennials-confident-connected-open-to-change (accessed October 7, 2013).

15. David A. Graham, "The Surreal Semiotics of Burning Obamacare Draft Cars," *The Atlantic,* August 2, 2013, http://www.theatlantic.com/politics/archive/2013/08/the-surreal-semiotics-of-burning-obamacare-draft-cards/278321 (accessed December 12, 2013).

16. Kevin Drum, "FreedomWorks Plans Push to Persuade People Not to Get Health Insurance," *Mother Jones,* July 25, 2013, http://www.motherjones.com/kevin-drum/2013/07/freedomworks-plans-push-persuade-people-not-get-health-insurance (accessed September 27, 2013).

17. Amanda Terkel, "Kathleen Sebelius Criticizes 'Dismal' Conservative Effort Urging Young People Not to Enroll in ObamaCare," *Huffington Post,* August 5, 2013, http://www.huffingtonpost.com/2013/08/05/kathleen-sebelius-obamacare_n_3708198.html (accessed August 27, 2013).

18. Carla Johnson, "ObamaCare National Marketing Campaign to Cost Nearly $700 Million," *Real Clear Politics,* July 25, 2013, http://www.realclearpolitics.com/articles/2013/07/25/obamacare_national_marketing_campaign_to_cost_nearly_700_million_119368.html (accessed October 3, 2013).

19. Christopher Weaver and Louise Radnofsky, "New Health-Care Law's Success Rests on the Young," *Wall Street Journal,* July 25, 2013, http://online.wsj.com/article/SB10001424127887324263404578613700273320428.html (accessed September 26, 2013).

20. David Hogberg, "Why the 'Young Invincibles' Won't Participate in the ObamaCare Exchanges and Why It Matters," National Center for Public Policy Research, August 2013, http://www.nationalcenter.org/NPA652.html (accessed September 26, 2013).

21. Louise Radnofsky, "Prices Set for New Health-Care Exchanges," *Wall Street Journal,* September 25, 2013, http://online.wsj.com/article/SB10001424052702303983904579095731139251304.html (accessed September 27, 2013).

22. Sam Cappellanti, "Premium Increases for 'Young Invincibles' Under the ACA and the Impending Premium Spiral," American Action Forum, October 2, 2013, http://americanactionforum.org/research/

premium-increases-for-young-invincibles-under-the-aca-and-the
-impending-premium-spiral (accessed October 3, 2013).

23.   Nick Gillespie, "Ads Hide Obamacare Truth: It's Generational Theft,"
      Time, November 25, 2013, http://content.time.com/time/magazine/
      article/0,9171,2157491,00.html (accessed December 5, 2013).

24.   Joel Stein, "Millennials: The Me Me Me Generation," *Time,* May 20, 2013,
      http://content.time.com/time/magazine/article/0,9171,2143001,00
      .html (accessed September 27, 2013).

25.   T. Scott Gross, *Invisible: How Millennials are Changing the Way We Sell*
      (Bloomington, IN: Triple Nickel Press, 2012).

26.   Leonard Downie, Jr., "The Obama Administration and the Press,"
      Committee to Protect Journalists, October 10, 2013, http://cpj.org/
      reports/2013/10/obama-and-the-press-us-leaks-surveillance-post-911
      .php (accessed December 15, 2013).

27.   "Survey of Young Americans' Attitudes Toward Politics and Public
      Service: 23rd Edition," Institute of Politics, Harvard University, April
      30, 2013, 14–15, http://www.iop.harvard.edu/sites/default/files_new/
      spring_poll_13_Exec_Summary.pdf (accessed August 26, 2013).

28.   "63% View Too-Powerful Government as Bigger Threat than Weaker
      One," *Rasmussen Reports,* July 3, 2013, http://www.rasmussenreports
      .com/public_content/politics/general_politics/june_2013/63_view_
      too_powerful_government_as_bigger_threat_than_weaker_one (ac-
      cessed August 22, 2013).

29.   Aamer Madhani, "Obama to Youth: Be Responsible and Sign up for
      Obamacare," *USA Today,* December 4, 2013, http://www.usatoday
      .com/story/news/politics/2013/12/04/obama-young-people-obliga
      tion-obamacare/3871351 (accessed December 6, 2013).

30.   Ron Fournier, "Millennials Abandon Obama and Obamacare," *Na-
      tional Journal,* December 4, 2013, http://www.nationaljournal.com/
      politics/millennials-abandon-obama-and-obamacare-20131204 (ac-
      cessed December 15, 2013).

31.   Nate Silver, "Poll Finds a Shift Towards More Libertarian Views,"
      New York Times, June 20, 2011, http://fivethirtyeight.blogs.nytimes
      .com/2011/06/20/poll-finds-a-shift-toward-more-libertarian-views/?_
      r=0 (accessed August 22, 2013).

32.   CNN Opinion Research Poll, June 17, 2013, http://i2.cdn.turner.com/
      cnn/2013/images/06/17/rel7a.pdf (accessed August 22, 2013).

33.   FreedomWorks, "The Role of Government."

34.   James Hohmann, "Poll: Republicans Embracing Libertarian Priorities,"
      Politico, September 11, 2013, http://www.politico.com/story/2013/09/
      poll-republicans-libertarian-96576.html (accessed September 12,
      2013).

35.   Rebekah Metzler, "Obama: I Am Not a Socialist," *U.S. News and*

World Report, November 19, 2013, http://www.usnews.com/news/articles/2013/11/19/obama-i-am-not-a-socialist (accessed December 15, 2013).

Chapter 6: The Right to Know

1. Adam Ferguson, "An Essay on the History of Civil Society" (1767), http://oll.libertyfund.org/?option=com_staticxt&staticfile=show.php%3Ftitle=1428&chapter=19736&layout=html#a_156617 (accessed November 6, 2013).

2. John Perry Barlow, Closing Keynote Address at the Electronic Frontier Foundation, June 8, 2013.

3. Deb Reichman, "Kerry: Some NSA Surveillance Reached 'Too Far,' " *U.S. News & World Report,* November 1, 2013 http://www.usnews.com/news/politics/articles/2013/11/01/kerry-some-nsa-surveillance-reached-too-far (accessed November 6, 2013).

4. Nat Hentoff, "Our Constitution: How Many of Us Know It?" *Cato,* May 19, 2011, http://www.cato.org/publications/commentary/our-constitution-how-many-us-know-it (accessed November 6, 2013).

5. Chris Moody, "Lindsey Graham: 'If I Thought Censoring the Mail Was Necessary, I Would Suggest It,' " *Yahoo News,* June 11, 2013, http://news.yahoo.com/blogs/ticket/lindsey-graham-thought-censoring-mail-necessary-suggest-182932835.html (accessed November 6, 2013).

6. Interview with Lindsey Graham, Fox News, June 6, 2013, http://www.youtube.com/watch?v=HzsxsTEk7tc (accessed November 6, 2013).

7. Glenn Greenwald, "NSA Collecting Phone Records of Millions of Verizon Customers Daily," *Guardian,* June 5, 2013, http://www.theguardian.com/world/2013/jun/06/nsa-phone-records-verizon-court-order (accessed August 28, 2013).

8. Barton Gellman and Laura Poitras, "U.S., British Intelligence Mining Data from Nine U.S. Internet Companies in Broad Secret Program," *Washington Post,* June 6, 2013, http://www.washingtonpost.com/investigations/us-intelligence-mining-data-from-nine-us-internet-companies-in-broad-secret-program/2013/06/06/3a0c0da8-cebf-11e2-8845-d970ccb04497_story.html (accessed August 28, 2013).

9. Glenn Greenwald and Ewen MacAskill, "NSA Prism Program Taps Into User Data of Apple, Google and Others," June 6, 2013, http://www.theguardian.com/world/2013/jun/06/us-tech-giants-nsa-data (accessed August 28, 2013).

10. Tom McCarthy, "Holder Ducks NSA Phone Record Questions in Senate Hearing—As It Happened," *Guardian,* June 6, 2013, http://www.theguardian.com/world/2013/jun/06/holder-phone-records-surveil

lance-live?utm_source=dlvr.it&utm_medium=twitter#block-51b0af76e 4b0cc64243720d3 (accessed August 28, 2013).

11. Meghashyam Mali and Brandon Sasso, "Administration Defends NSA Grab of Verizon Customer Phone Calls," *The Hill,* June 6, 2013, http:// thehill.com/blogs/hillicon-valley/technology/303821-white-house -defends-nsa-collecting-verizon-phone-records (accessed August 28, 2013).

12. Andrew Rosenthal, "Making Alberto Gonzales Look Good," *New York Times,* June 11, 2013, http://takingnote.blogs.nytimes.com/ 2013/06/11/making-alberto-gonzales-look-good/?_r=0 (accessed August 28, 2013).

13. Michael Pearson, "Obama: No One Listening to Your Calls," CNN, June 10, 2013, http://edition.cnn.com/2013/06/07/politics/nsa-data -mining (accessed August 28, 2013).

14. Mike Dorning and Chris Strohm, "Secret Court Finding Domestic Spying Risks Obama Credibility," Bloomberg News, August 23, 2013, http://www.businessweek.com/news/2013-08-23/secret-court-finding -domestic-spying-risks-obama-credibility (accessed August 28, 2013).

15. Siobhan Gorman, "NSA Officers Spy on Love Interests," *Wall Street Journal,* August 23, 2013, http://realclearscience.com/blog/2013/08/ how-can-americans-be-both-obese-and-starving.html (accessed August 27, 2013).

16. "President Obama's Dragnet," *New York Times,* June 6, 2013, http:// www.nytimes.com/2013/06/07/opinion/president-obamas-dragnet .html?pagewanted=all&_r=0 (accessed August 28, 2013).

17. Gerald F. Seib, "In Crisis, Opportunity for Obama," *Wall Street Journal,* November 21, 2008, http://online.wsj.com/article/SB12272127 8056345271.html (accessed September 9, 2013).

18. Liz Klimas, "Was Watertown's Door-to-Door Search for Bombing Suspects a Violation of the Fourth Amendment?," *The Blaze,* April 13, 2013, http://www.theblaze.com/stories/2013/04/23/ready-how-water town-door-to-door-search-for-bombing-suspects-did-not-violate-the -fourth-amendment (accessed October 10, 2013.)

19. "Obama on NSA Surveillance: Can't Have 100% Security and 100% Privacy," RT.com, June 7, 2013, http://rt.com/usa/obama-surveillance -nsa-monitoring-385 (accessed September 9, 2013).

20. Interview with Representative Jim Sensenbrenner, NPR, August 20, 2013, http://www.npr.org/templates/story/story.php?storyId=213902 177 (accessed October 10, 2013).

21. Andrew Napolitano, "Government Spying Out of Control," *Reason,* December 13, 2012, http://reason.com/archives/2012/12/13/govern ment-spying-out-of-control (accessed October 15, 2013).

22. Glenn Greenwald and James Ball, "The Top Secret Rules that Allow

NSA to Use US Data without a Warrant," *Guardian,* June 20, 2013, http://www.theguardian.com/world/2013/jun/20/fisa-court-nsa-with out-warrant (accessed October 10, 2013).

23. Office of the Attorney General, Letter to the Honorable Rand Paul, March 4, 2013, http://www.paul.senate.gov/files/documents/Brennan HolderResponse.pdf (accessed November 6, 2013).

24. Morgan Little, "Transcript: Rand Paul's Filibuster of John Brennan's CIA Nomination," *Los Angeles Times,* March 7, 2013, http://articles .latimes.com/2013/mar/07/news/la-pn-transcript-rand-paul-filibuster -20130307 (accessed October 29, 2013).

25. Transcript of "The Last Word with Lawrence O'Donnell," March 7, 2013, http://livedash.ark.com/transcript/the_last_word_with_lawrence _o%27donnell/5304/MSNBC/Thursday_March_07_2013/619583;sh (accessed December 15, 2013).

26. "McCain Slams Rand Paul for Filibuster: 'Calm Down, Senator,' " *Real Clear Politics* Video, March 7, 2013, http://www.realclearpolitics .com/video/2013/03/07/mccain_slams_rand_paul_for_filibuster_ calm_down_senator.html (accessed December 15, 2013).

27. Stephen Dinan, "Graham, McCain Blast Paul Filibuster," *The Washington Times,* March 7, 2013, http://www.washingtontimes.com/blog/ inside-politics/2013/mar/7/graham-mccain-blast-paul-filibuster/#ixzz 2nTAqgPDn (accessed December 15, 2013).

28. Timothy Noah, "The Legend of Strom's Remorse," *Slate,* December 16, 2002, http://www.slate.com/articles/news_and_politics/chatter box/2002/12/the_legend_of_stroms_remorse.html (accessed September 10, 2013).

29. *United States Congressional Record,* August 28, 1957, 16402, http://www .archive.org/stream/congressionalrec103funit#page/n1061/mode/ 1up (accessed September 11, 2013).

30. "Thurmond Holds Senate Record for Filibustering," Associated Press, June 27, 2013, http://www.foxnews.com/story/0,2933,90552,00.html (accessed August 21, 2013).

31. Patrick Ruffini, "#StandWithRand Blowing Up Right Now," Twitter, March 6, 2013, https://twitter.com/PatrickRuffini/status/30950464 5728980995/photo (accessed September 18, 2013).

32. TrendPo, "StandWithRand," August 2, 2013, http://blog.trendpo .com/2013/08/02/download-trendpos-latest-white-paper-standwith rand (accessed September 18, 2013).

33. Josh Vorhees, "Rand Paul Ends Epic Mr. Smith–Style Filibuster After More Than 12 Hours," *Slate,* March 7, 2013, http://www.slate.com/ blogs/the_slatest/2013/03/06/rand_paul_is_waging_an_epic_mr_ smith_style_filibuster_right_ now.html (accessed August 21, 2013).

34. Partisan ID, "Rand Paul's 2016 GOP Primary Poll Bounce Has Ar-

rived, Courtesy of an Old Fashioned Filibuster," April 4, 2013, http://
partisanid.blogspot.com/2013/04/rand-pauls-2016-gop-primary-poll
-bounce.html (accessed November 6, 2013).

35. Mo Elleithee, "New Leader of the GOP: Rand Paul," CNN, March 8,
2013, http://www.cnn.com/2013/03/08/opinion/elleithee-gop-rand
-paul/ (accessed November 6, 2013).

36. Brett Logiurato, "Since Rand Paul's Historic Filibuster, There Has
Been a Dramatic Shift in Public Opinion on Drone Strikes," *Business
Insider,* April 11, 2013, http://www.businessinsider.com/rand-paul-fili
buster-drone-polling-polls-2013-4 (accessed December 5, 2013).

37. Office of the Attorney General, Letter to the Honorable Rand Paul,
March 7, 2013, http://www.washingtonpost.com/blogs/post-politics/
files/2013/03/Senator-Rand-Paul-Letter.pdf (accessed November 6,
2013).

38. Brett Molina, "Five Questions About the Business of Twitter," *USA
Today,* October 3, 2013, http://www.usatoday.com/story/tech/2013/
10/03/five-questions-twitter-s1/2918277/ (accessed October 4, 2013).

39. Sydney Brownstone, "Twitter vs. Mainstream Media: Science Proves
Which Breaks News Faster," Fastcoexist.com, July 9, 2013, http://
www.fastcoexist.com/1682521/twitter-vs-mainstream-media-science
-proves-which-breaks-news-faster (accessed October 4, 2013).

40. Roy Morejon, "How Social Media is Replacing Traditional Journalism
as a News Source," *Social Media Today,* June 28, 2012, http://social
mediatoday.com/roymorejon/567751/how-social-media-replacing
-traditional-journalism-news-source (accessed August 22, 2013).

41. John Perry Barlow, Closing Keynote Address at the Electronic Frontier
Foundation, June 8, 2013.

42. Jim Newell, "Thousand Gather in Washington for Anti-NSA 'Stop
Watching Us' Rally," *Guardian,* October 26, 2013, http://www
.theguardian.com/world/2013/oct/26/nsa-rally-stop-watching
-washington-snowden (accessed November 6, 2013).

Chapter 8: Twelve Steps

1. "David Brooks Warns About 'the Rise of Ted Cruz-ism,' " *Real Clear
Politics,* video, September 15, 2013, http://www.realclearpolitics.com/
video/2013/09/14/david_brooks_warns_about_the_rise_of_ted_
cruz-ism.html (accessed September 30, 2013).

2. Daniel Politi, "Harry Reid: 'The American People Will Not Be Ex-
torted by Tea Party Anarchists,' " *Slate,* September 28, 2013, http://
www.slate.com/blogs/the_slatest/2013/09/28/harry_reid_on_shut

down_american_people_will_not_be_extorted_by_tea_party.html (accessed September 30, 2013).

3. Sam Baker, "Baucus Warns of 'Huge Train Wreck' Enacting ObamaCare Provisions," *The Hill,* April 17, 2013, http://thehill.com/blogs/healthwatch/health-reform-implementation/294501-baucus-warns-of-huge-train-wreck-in-obamacare-implementation (accessed September 30, 2013).

4. Barbara Hagenbaugh and Sue Kirchhoff, "Timothy Geithner Says He Regrets Tax Mistakes," *USA Today,* January 22, 2009, http://usatoday30.usatoday.com/money/economy/2009-01-21-geithner-hearing_N.htm (accessed October 9, 2013).

5. *Newsmax,* "60 Minutes Uncovers Pelosi's Insider Stock Trades," November 13, 2011, http://www.newsmax.com/insidecover/pelosi-stock-insider-60minutes/2011/11/13/id/417848 (accessed December 5, 2013).

6. S. 2038 (112th): STOCK Act, http://www.govtrack.us/congress/bills/112/s2038 (accessed September 23, 2013).

7. Stephen Dinan, "Congress Votes to Shield Top Officials' Financial Disclosures," *Washington Times,* April 12, 2013, http://www.washingtontimes.com/news/2013/apr/12/senate-votes-shield-top-officials-financial-disclo/?page=1 (accessed September 30, 2013).

8. Elise Viebeck, "GOP Bill Hits Alleged ObamaCare Exemption Talks," *The Hill,* April 26, 2013, http://thehill.com/blogs/healthwatch/health-reform-implementation/296489-gop-bill-hits-alleged-obamacare-exemption-talks#ixzz2RbQ3O72J (accessed September 23, 2013).

9. Matt Kibbe, *Hostile Takeover* (New York: HarperCollins, 2012), 1.

10. S. J. Res. 1 (113th): Proposing an Amendment to the Constitution of the United States Requiring That the Federal Budget Be Balanced, http://thomas.loc.gov/cgi-bin/query/z?c113:S.J.RES.1.IS: (accessed September 24, 2013).

11. Congressional Budget Office, "The 2013 Long-Term Budget Outlook," http://www.cbo.gov/publication/44521 (accessed September 27, 2013).

12. "Sen. Ted Cruz Leading the Charge on Abolishing the IRS," Fox News, video, June 3, 2013, http://video.foxnews.com/v/2429567001001/sen-ted-cruz-leading-the-charge-on-abolishing-the-irs (accessed September 23, 2013).

13. National Center for Policy Analysis, "The Medicare/Social Security Trustees Spring Report: A Bleak Future," http://www.ncpa.org/pdfs/A_Bleak_Future.pdf (accessed September 30, 2013).

14. Joseph Lawler, "Economist Laurence Kotlikoff: U.S. $222 Trillion in Debt," *Real Clear Policy,* December 1, 2013, http://www.realclearpolicy.com/blog/2012/12/01/economist_laurence_kotlikoff_us_222_trillion_in_debt_363.html (accessed September 27, 2013).

15. Rand Paul, *A Clear Vision to Revitalize America,* March 22, 2013, http://www.paul.senate.gov/files/documents/FY2014Budget.pdf (accessed September 23, 2013).

16. S. 2196 (112th): Congressional Health Care for Seniors Act, http://www.govtrack.us/congress/bills/112/s2196# (accessed September 23, 2013).

17. Peter Suderman, "Rand Paul, Jim DeMint, and Mike Lee's Medicare Plan Is a Challenge to Both Sides of the Health Care Debate," *Reason,* March 16, 2012, http://reason.com/blog/2012/03/16/rand-paul-jim-demint-and-mike-lees-medic (accessed September 23, 2013).

18. "Transcript of Speaker Pelosi's Speech," *New York Times,* September 29, 2008, http://www.nytimes.com/2008/09/30/washington/30pelosi transcript.html?pagewanted=all (accessed September 30, 2013).

19. "Hensarling Re-Introduces Legislation to End Taxpayer Funded Bailout of Fannie Mae and Freddie Mac," March 18, 2011, http://hensarling .house.gov/news/press-releases/2011/03/hensarling-re-introduces -legislation-to-end-taxpayer-funded-bailout-of-fannie-mae-and-freddie -mac.shtml (accessed September 23, 2013).

20. Press Release, "Congressmen Write Landmark Surveillance Reform," *Office of Congressman Justin Amash,* October 29, 2013, http://amash .house.gov/press-release/congressmen-write-landmark-surveillance -reform (accessed December 6, 2013).

21. Katrina Trinko, "The Phone Lines Melt," *National Review,* September 6, 2013, http://www.nationalreview.com/article/357771/phone-lines -melt-katrina-trinko (accessed December 15, 2013).

22. CNN Wire Staff, "Mullen: Debt Is Top National Security Threat," *CNN,* August 27, 2010, http://www.cnn.com/2010/US/08/27/debt .security.mullen/index.html?_s=PM:US (accessed December 5, 2013).

23. *Susette Kelo, et al., Petitioners v. City of New London, Connecticut, et al.,* 04-108 U.S. (2005), available at http://caselaw.lp.findlaw.com/scripts/ getcase.pl?court=us&vol=000&invol=04-108 (accessed September 30, 2013).

24. Bob Ewing, "Taken: Federal Lawsuit in Michigan Challenges Forfeiture Abuse," Institute for Justice, September 25, 2013, http://www .ij.org/michigan-civil-forfeiture-release-9-25-2013 (accessed October 1, 2013).

25. Joel Brinkley, "What Tyrants Fear Most: Social Media," The *Chicago Tribune,* November 27, 2012, http://articles.chicagotribune .com/2012-11-27/opinion/sns-201211271500—tms—amvoicesctnav -c20121127-20121127_1_social-media-bloggers-president-alexander -lukashenko (accessed December 16, 2013).

26. Scott Gottlieb, "Suit Alleges IRS Improperly Seized 60 Million Per-

sonal Medical Records," *Forbes,* May 15, 2013, http://www.forbes
.com/sites/scottgottlieb/2013/05/15/the-irs-raids-60-million-personal
-medical-records (accessed October 3, 2013).

27. Declan McCullagh, "IRS Claims It Can Read Your E-mail without
a Warrant," CNET, April 10, 2013, http://news.cnet.com/8301-13578
_3-57578839-38/irs-claims-it-can-read-your-e-mail-without-a-warrant/
(accessed October 3, 2013).

28. See United States Copyright Office, Circular 92, "Copyright Law of
the United States of America and Related Laws Contained in Title 17
of the *United States Code,*" http://www.copyright.gov/title17/92chap3
.html#302 (accessed October 10, 2013).

29. Mark M. Jaycox and Kurt Opsahl, "CISPA Is Back: FAQ on What It Is
and Why It's Still Dangerous," *Electronic Future Foundation,* February
25, 2013, https://www.eff.org/cybersecurity-bill-faq (accessed December 16, 2013).

Chapter 9: Not a One-Night Stand

1. George Jean Nathan and H. L. Mencken, "Repetition Generale," *The
Smart Set,* December 1919, p. 71.

2. Jill Lawrence, "Former House Leaders Say the Current Group Has It
Rough," *National Journal,* September 23, 2013, http://www.national
journal.com/daily/former-house-leaders-say-the-current-group-has-it
-rough-20130923 (accessed November 4, 2013).

3. Peter Schweizer, *Throw Them All Out* (Boston: Houghton Mifflin Har-
court, 2011).

4. Grace Wyler and Zeke Miller, "Beyond Insider Training: Here's How
Members of Congress Get Rich off Earmarks," November 15, 2011,
http://www.businessinsider.com/congress-insider-trading-earmarks
-real-estate-nancy-pelosi-rich-tax-payer-money-2011-11?op=1#ixzz2
jPlZMcXW (accessed November 4, 2013).

5. Michael Gerson, " 'Purity for Profit' Wing Threatens GOP's Stability," *The Daytona Beach News-Journal,* November 7, 2013, http://www
.news-journalonline.com/article/20131107/WIRE/131109649   (accessed December 5, 2013).

6. Matthew Yglesias, "It's Worth Actually Reading Obama's 2006 Debt
Ceiling Speech," *Slate,* October 9, 2013, http://www.slate.com/blogs/
moneybox/2013/10/09/obama_s_2006_debt_ceiling_speech.html
(accessed November 4, 2013).

7. Jeffrey Lord, "Cruz Leads the Reaganite Rebellion," *American Spectator,* September 24, 2013, http://spectator.org/archives/2013/09/24/
cruz-leads-the-reaganite-rebel/print (accessed November 4, 2013).

8. Manuel Klausner, "Inside Ronald Reagan," *Reason,* July 1975, http://reason.com/archives/1975/07/01/inside-ronald-reagan (accessed November 4, 2013).
9. Benjamin Franklin in a pseudonymous letter published in the *Pennsylvania Journal* on December 27, 1775.

INDEX

Photo by Sam Hurd

About the Author

Matt Kibbe is the president and CEO of FreedomWorks, a national grassroots organization that serves citizens in their fight for more individual freedom and less government control. An economist by training, Kibbe is a well-respected policy expert, bestselling author, and regular guest on CNN, Fox News, The Blaze TV, and MSNBC. He also serves as Distinguished Senior Fellow at the Austrian Economic Center in Vienna, Austria. Kibbe is the author of the national bestseller *Hostile Takeover: Resisting Centralized Government's Stranglehold on America* (2012) and coauthor of *Give Us Liberty: A Tea Party Manifesto* (2010). Terry, his awesome wife of twenty-eight years, takes no responsibility for his many mistakes or frequent embarrassments.

BOOKS BY MATT KIBBE

DON'T HURT PEOPLE AND DON'T TAKE THEIR STUFF
A Libertarian Manifesto

Available in Paperback and eBook

Don't Hurt People and Don't Take Their Stuff is a rational yet deeply felt argument supporting the individual against the oppression of the collective. Matt Kibbe points out that the political and corporate establishment in Washington consolidates its power by infringing upon basic American freedoms. Libertarianism is one of the hottest political movements in contemporary America, animating both the right and left with its emphasis on personal liberty. Espoused everywhere from young activists on college campuses to the stars of the right wing like Marco Rubio and Rand Paul, libertarian ideology is gripping frustrated conservatives and political contrarians.

HOSTILE TAKEOVER
Resisting Centralized Government's Stranglehold on America

Available in Paperback and eBook

A rebellious challenge to the "upper management" of government, who are choking American prosperity and liberty. Matt Kibbe, the high-profile leader of FreedomWorks—who Geraldo Rivera calls "a warrior for the cause of limited government"—now offers an intelligent, aggressively argued attack on the American federal government machine in Washington, D.C. *In Hostile Takeover*, Kibbe, co-author of the #1 bestseller *Give Us Liberty*, provides a blueprint for "resisting centralized government's stranglehold on America," in order to return the nation to the more workable system our Founding Fathers originally intended.